High **Technē**

Electronic Mediations

Katherine Hayles, Mark Poster, and Samuel Weber, series editors

Volume 2. *High Technē: Art and Technology from the Machine Aesthetic to the Posthuman*
R. L. Rutsky

Volume 1. *Digital Sensations: Space, Identity, and Embodiment in Virtual Reality*
Ken Hillis

Electronic Mediations, volume 2

Posthuman
Art and
Technology
from the
Machine
Aesthetic
to the

High Technē

Art and Technology from the Machine Aesthetic to the Posthuman

Art and
Technology
from the
Machine
Aesthetic
to the
Posthuman
Art and
Technology
from the
Machine
Aesthetic
to the
Posthuman

R. L. Rutsky

University of Minnesota Press
Minneapolis • London

Chapter 2 was previously published as "The Mediation of Technology and Gender: *Metropolis,* Nazism, Modernism," *New German Critique* 60 (fall 1993); reprinted with permission.

Published by the University of Minnesota Press
111 Third Avenue South, Suite 290
Minneapolis, MN 55401-2520
http://www.upress.umn.edu

Library of Congress Cataloging-in-Publication Data

Rutsky, R. L.
High technē : art and technology from the machine aesthetic to the posthuman / R. L. Rutsky.
p. cm. — (Electronic mediations ; v. 2)
Includes index.
ISBN 0-8166-3355-X (HC). — ISBN 0-8166-3356-8 (PB)
1. Art and technology—Philosophy. 2. Technology—Aesthetics.
I. Title. II. Series.
N72.T4R87 1999
701'.05—dc21 99-31609

Printed in the United States of America on acid-free paper

11 10 09 08 07 06 05 04 03 02 01 10 9 8 7 6 5 4 3 2

Contents

Introduction
The Question concerning High Tech

From the perspective of the ever more technologized cultures of the industrialized world, it seems increasingly difficult to avoid the sense that, somehow, the entire world has undergone an indefinable but undeniable change, a kind of mutation. Thus, for example, Jean Baudrillard can speak of "the mutation of [a] properly industrial society into what could be called our techno-culture."[1] This sense of a techno-cultural mutation has, of course, frequently been figured in terms of postmodernity—as part of a broader shift from modern to postmodern. But then, the very notions of both modernity and postmodernity are quite simply inconceivable without technology. This is not to say, however, that technology is the "determining instance" of either modern or postmodern culture, nor that the current sense of a techno-cultural mutation is based on particular changes in technology. Rather, whatever changes or mutations have occurred in contemporary cultures—whether one calls these cultures postmodern or not—seem to be based less on changes in technology per se than in the very conception of technology, of what technology is.

There have, of course, been innumerable discussions of technology and the interrelation of technological and cultural change in recent years. Popular magazines such as *Time* and *Newsweek* have featured cover stories on "cyberpunk" and "techno-mania," and other magazines devoted entirely to "new tech" and "high tech"—such as *Wired, Mondo 2000,* and *Boing, Boing*—have sprung into existence. Nor have university presses and academic journals ignored the issue of technological and cultural change, even if their discussions have often taken place under the somewhat broader rubric of "postmodern culture" or "techno-culture."

Yet, despite the time and energy devoted to the issue, the debates over technology and techno-culture often seem to have a wearisome sameness.

Even when the debate concerns technological change, the terms of the debate do not seem to change at all: technology—or some aspect of it—is either celebrated or decried, cast as utopian or dystopian, in terms of its capacity either to serve "humanity" or to threaten it. The repetitiveness of these arguments results from the fact that they take the *definition* of technology for granted. For all the discussion of the implications of technological change, remarkably little attention has been devoted to possible changes in the *conception* of technology.

In other words, even as views of technology have—in an age of high technology—implicitly changed, the definition of technology has remained largely unquestioned. What has been left unexamined, then, is precisely Heidegger's "question concerning technology"—which is not, for Heidegger, a question of technology per se, but rather of what he calls "the essence *[Wesen]* of technology," which "is by no means anything technological."[2] Indeed, Heidegger argues that it is just this nontechnological "essence" that has been obscured by the commonly accepted definition of technology as instrumental, as a means to an end. For Heidegger, this instrumental conception of technology—although it presumes to define "what technology is," to define "the technological"—is merely the modern manifestation of "the essence of technology." In other words, the modern conception of technology, because it restricts the definition of the technological to instrumental terms, "blinds us to" that broader "essence" that informs not only the modern view of technology, but also the quite different conceptions of traditional technology and the *technē* of ancient Greece. Thus, Heidegger seeks to reenvision not only what technology is, but what it can be. Heidegger's "broader" view of technology, therefore, seems particularly appropriate to the question of how the conception of technology may have changed in an age of high technology—appropriate, that is, to what might well be called "the question concerning high tech."

This question concerning high tech is, as Heidegger suggests, a historical question. The very notion of modernity—from its beginnings in the Renaissance's image of itself as a new age, a historical break from the "Dark Ages"—has been defined in terms of an instrumental conception of technology, an instrumental or technological rationality that allows modern "humanity" to know and control the world. In these terms, that which is "nontechnological" cannot be modern.[3] If, however, Heidegger questions the "universality" of this instrumental conception of technology by pointing to its historical specificity (as modern), he neglects the extent to which it is also culturally specific. Modernity, defined in terms of an instrumental technology and rationality, has long been the basis on which Western, patriarchal cultures have privileged themselves over their "non-technological" others.[4] From this perspective, cultures or discourses—for example, "non-Western" cultures, "feminine" discourses—that perceive the world in terms other than those of rational, scientific knowledge and technological control are necessarily characterized as antimodern, irrational, often even as "primitive." Thus, although the sense of a cultural, technological mutation may itself be specific to "highly technologized" cultures, its implications are not; for, if in high technology the modern conception of technology has changed, so too has the relation of "techno-culture" to those supposedly nontechnological "other" cultures and discourses that modernity has always devalued, excluded, or repressed.

High Technē

"High technology" would seem, at first glance, to be simply a matter of *more* technology—that is, a more extreme, more effective version of modern technology. And certainly, the instrumental or functional conception that defines modern technology remains an important aspect of high technology, or "high tech." No one could deny the uncanny "functionality" of those military and "Star Wars" technologies that have allowed war and killing to be instrumentalized to an unprecedented degree. Nor is it possible to disregard the efficiency with which various information technologies enable an increasing differential, in terms of both economic and knowledge capital, between the technologically rich and the technologically poor. In this sense, high technology continues to maintain a distinction between a "high" and a "low" culture, between those who have a "high" level of access to technology and those who do not. Thus, despite the pronouncements of various technological "visionaries" and corporate chiefs detailing how "high tech" will "democratize" society, enabling universal access, participation, and control over one's life, high technology remains a "tool" for distinguishing social classes.

Yet, at the same time, high tech also involves—and indeed, seems to highlight—a noninstrumental or "nontechnological" aspect that, as Heidegger observes, has been largely obscured in the modern conception of technology. In fact, this "nontechnological" aspect—crucial both to Heidegger's "essence of technology" and to "high tech"—is linked to a realm that has generally been cast as the polar opposite of modern technology: that of art and aesthetics.

From its very beginnings, in fact, the conception of technology in Western culture seems to have been defined by its shifting relationship to the realm of art. Thus, for example, Heidegger finds that in the Greek root of technology, *technē*—generally translated as "art," "skill," or "craft"—technology and art were closely linked. For the Greeks, "it was not technology alone that bore the name *technē*," but art too "was simply called *technē*" (*TQCT,* p. 34). Heidegger's point, however, is not that technology's close relationship to art in ancient Greece has simply been lost. Rather, he argues that the relationship between art and technology, so visible in the Greek *technē*, has always been basic to technology, to its "essence," even when the *conception* of technology has been explicitly posed (as it has in the modern, instrumental conception of technology) in contrast to art, to the aesthetic sphere. High tech, with its emphasis on issues of representation, style, and design, seems to signal a reemergence of this repressed aesthetic aspect within the conception of technology.

Unlike modern technology, high tech can no longer be defined *solely* in terms of its instrumentality or function—as simply a tool or a means to an end. In high tech, rather, technology becomes much more a matter of representation, of aesthetics, of style. This concern with representation and style displays itself not only in the design of technological objects themselves, but also in the practice of imparting a "high-tech look" or style to objects that are not in themselves highly technological. Thus, items as various as basketball shoes and exposed pipes and ducts have been described as having a "high-tech style." In "high-tech design," then, the modernist ideal of functional form has been largely abandoned in favor of a technological look or style that need not be functional in any traditional sense; the efficacy of such items becomes, for the most part, a matter of cultural style, cultural desires. Yet, the high-tech concern with style and stylishness is not limited to questions of design; in high tech, the very "function" of technology becomes a matter of representation, style, aesthetics—a matter, that is, of technological reproducibility. In high tech, the ability to technologically reproduce, modify, and reassemble stylistic or cultural elements becomes not merely a means to an end, but an end in itself. This process of technological representation, of reproducibility, alteration, and assemblage, can be said to define high tech. High technology is simulacral technology: a technology "of reproduction rather than of production," as Fredric Jameson has said of late-capitalist or postmodern technologies.[5] What this technology reproduces—and thus puts "into play"—is representation itself, style itself. But then, representation and style have always been technological, supplementary, simulacral. In high tech, however, this simulacral status becomes an end in itself, rather than merely a means to an end or a copy of an original.

To speak of a high-tech aesthetic or style is not, then, simply to speak of a particular look or style, but of a cultural concern with "stylishness," with "aesthetics," that is intrinsic to high tech. Indeed, high tech is by definition a technology that is "at the state of the art in terms of . . . function and design."[6] To be "at the state of the art" implies not only a certain up-to-the-second currency, an attention to the latest technological developments, but also a sense that both "function and design" have become elements in an aesthetic process or movement. As state-of-the-art technology, high tech comes to be defined by its status as the "cutting edge" or "leading wave" of this technological aesthetic or style. Indeed, it is no coincidence that the often overblown rhetoric associated with high tech is reminiscent of the manifestos and slogans of the avant-garde artistic movements of the early twentieth century. High tech is, in fact, often presented as a kind of avant-garde movement.

There are, of course, good reasons to be extremely skeptical of the "avant-garde" rhetoric of high tech (as there are, for that matter, of the rhetoric of the modernist avant-gardes). If the rhetoric of the modernist avant-gardes served to distinguish an artistic vanguard from the rest of the population, the notion of a high-tech avant-garde privileges a "highly technological" vanguard that is also, often, "highly capitalist." Yet, one crucial similarity that high tech does share with the modernist avant-gardes is that in both, the conjunction of the technological and the aesthetic is a central concern. Moreover, the very fact that metaphors such as "state of the art" and "avant-garde" have been so commonly employed—and accepted—in describing high tech is evidence that an "aesthetic" dimension has become part of the definition of contemporary technology. Technology has come increasingly to be seen as a matter of aesthetics or style, as an "aesthetic movement." Given this "aesthetic" aspect, the concept of technology in high tech might well be thought of as a kind of *high technē*—analogous to, though certainly quite different from, the Greek notion of *technē*.

The question, then, of how the modern conception of technology has changed to a *high technē* will necessarily involve charting the vicissitudes, the history, of the relationship between the technological and the aesthetic in modernity and beyond. Charting that history is, in fact, the project of this book. The book is, therefore, divided into two sections of two chapters each, with a transitional chapter between them: the first two chapters deal with the beginnings of this shift in the conception of technology and of aesthetics, concentrating on relations of art and technology in artistic modernism; the last two chapters focus on high tech itself, and on how the conceptions both of technology and of the aesthetic have changed in contemporary times. This project is not, however, simply a matter of describing *what* has changed, of comparing "snapshots" of the conception of technology "before and after" an epistemic, postmodern break. Indeed, this book suggests that this shift should be seen neither as revolutionary nor as simply evolutionary, but instead as an "emergent" process, in which a complex interaction of factors leads to a major change. To this end, the middle chapter (chapter 3) focuses on the modernist avant-gardes as a crucial transitional phase—and, as noted earlier, one that is still quite relevant—in the emergence of high tech out of modernist technological aesthetics. Throughout the book, in fact, the issue of *how* these changes in the conception of technology and the aesthetic took place has been emphasized over the question of *what* has changed: what, for example, was there in the modern conception of technology—and in its relation to aesthetics—that enabled the concept of high tech to develop out of it? In this regard, Heidegger's notion of the "essence of technology" seems to offer a suggestive way to conceptualize how this shift comes about, prior to the more detailed discussion of the movement from modern technology to high tech that will emerge in the chapters that follow.

The Turning of Technology

For Heidegger, this "essence of technology" cannot simply be defined in terms of the usual, modern sense of technology as an instrument, tool, or machine. He attempts instead to broaden the notion of technology into a more general concept of making or producing, including artistic production. The "essence of technology" is therefore not a static conceptual category or ideal, but a dynamic, ongoing process or movement. Thus, for example, Samuel Weber can translate Heidegger's *Wesen* as "goings-on," and can note that "As something that goes on, technics moves away from itself in being what it is."[7] In other words, the essence of technology is a matter of an ongoing change or movement that Heidegger refers to as *Entbergung*, a term that is usually translated as "revealing" or "disclosure." Yet, as Weber argues, *Entbergung* might also be translated as "unsecuring," because it also carries the implication of "a dismantling, an unleashing or releasing of an ambiguous, indeed highly conflictual dynamic."[8] As a form of *Entbergung*, then, technology has always been an ambiguous movement or process. It involves a "setting up" or "setting forth" that brings things into representation, sets them in place, in order. Yet, this setting in place or into representation can only "take place" inasmuch as technology is, at the same time, an unsettling movement or change (as in "setting forth" on a journey): an unsecuring that breaks things free and brings them forth, into representation, into play.[9]

As the essence of technology, this ambiguous representation or setting forth is present in both modern technology and the Greek sense of *technē*. Yet, the *mode* of this setting forth does change. Thus, for Heidegger, the mode of representation involved in the Greek *technē* is a form of unsecuring that is noninstrumental, and thus more closely related to artistic production *(poiesis)* than to the production of modern technology, which regulates and secures the world in instrumental terms. The world is thus "set in place" *(gestellt)*, which is why Heidegger figures the essence of modern technology, its mode of representation, as a kind of Enframing *[Ge-stell]*. Thus, while Enframing stresses setting in place, regulating, and securing, the emphasis in technē is on setting free, on unsecuring, on allowing the world to be "brought forth" in noninstrumental terms.

For Heidegger, the history of modernity can be read as an ever-increasing technological effort to regulate and secure the unsettling, "artistic" aspects inherent in technē. Through this Enframing, the unsecuring tendency of technology is given a set destination, directed toward instrumental ends.[10] Indeed, humanity is itself subjected to this kind of instrumental ordering. Yet, the unsecuring tendency of technology does not simply fade from the scene, but remains within Enframing.

Although Heidegger would have no doubt objected to such a comparison, his notion of an "unsecuring" aspect within Enframing is strangely similar to the role of "mechanical reproduction" within modern technology. Both Heidegger's concept of unsecuring and Walter Benjamin's idea of mechanical reproduction—or, translating his German more precisely, technological reproducibility—are "dismantling" processes. Just as technological reproducibility breaks down the "enchantment" or "aura" of the aesthetic realm, allowing art to become functional, unsecuring allows a mythic or "enchanted" view of the world to be broken down and thus transforms the world into objects that are available for human use and control. Moreover, as Samuel Weber has noted, the process of unsecuring actually serves as a motive force for modern technology's attempts to control and secure the world in instrumental terms: "The effort [to establish control and security] is all the more 'frantic' or 'furious' *(rasend)* because it is constantly goaded on by the unsecuring tendency of technics as such."[11] Technological reproducibility seems to work in a similar way: although it involves an attempt to extend an instrumental rationality to the realm of art, in so doing, it produces a proliferation of images and data that have been broken free of any set meaning or context and that therefore require increasing efforts to resecure them in instrumental terms. Paradoxically, however, the extension of an instrumental rationality also extends the unsecuring tendency of technological reproducibility. In a high-tech world, then, the proliferation of technological reproducibility begins to outstrip the ability to resecure it. Here, technological reproducibility becomes an end in itself, no longer governed by an instrumental rationality, but only by its own reproductive logic, its own "aesthetic."

Despite his obvious discomfort with technological reproduction, Heidegger may have intuited this possibility of the modern, instrumental conception of technology reaching a point where it begins to undermine itself. In an often-ignored passage, he notes the "astounding possibility" that "the frenziedness of technology may entrench itself everywhere to such an extent that someday, throughout everything technological, the essence of technology may come to presence in the coming-to-pass of truth" (*TQCT*, p. 35). Thus, the extension of Enframing may itself lead to the "coming-to-pass" of a conception of technology more in keeping with the unsettling artistic "essence" that remains, ongoing, within it. This emergence of a different conception of technology out of the older modern notion cannot, therefore, be seen simply as a break; it might better be described, to use a Heideggerian term, as a "turning."

Here, however, it is not only the conception of technology that has changed, but also the notion of aesthetics. The aesthetic can no longer be figured in the traditional terms of aura and wholeness, nor in the modernist terms of instrumentality or functionality. Like technology, it too comes to be seen as an unsettling, generative process, which continually breaks elements free of their previous context and recombines them in different ways. In this way, the technological and the aesthetic begin to "turn" into one another. And although this coming together of technology and art may be very different from what Heidegger had in mind in his notion of technē, it still seems appropriate to refer to it as a high technē.

Modernist Aesthetics: The Aesthetic Turn

This "aesthetic turn" in the conception of technology does not, however, begin only with the inception of "high tech." Its beginnings can readily be seen in that strange conjunction of the technological and the aesthetic that occurs in the modernist aesthetics of the late nineteenth and early twentieth centuries. In fact, modernist aesthetics has often been defined precisely in terms of its relation to technology. Yet the relation between technology and the aesthetic within artistic modernism is a complex one, and cannot be defined, as it all too often is, simply in terms of a tendency toward "functional form" or a "machine aesthetic." Any consideration of the technological and the aesthetic within modernism must take account not only of this tendency to "technologize aesthetics," but also of the opposite tendency, as evidenced in the Nazi desire to "aestheticize politics"—or, more precisely, to aestheticize modern technological society. Nor can we ignore—especially given Heidegger's associations with National Socialism—the extent to which his notion of an aesthetic turning in technology is implicated in the Nazi vision of an aestheticized technological state. Yet the question here is not simply Heidegger's relation to fascism, but that of modernism more generally. For within modernism, the desire to "aestheticize technology" is not limited to those who express explicitly reactionary or fascistic political sentiments. Indeed, it occurs with such regularity—even among the left avant-gardes—that it must be considered as much a part of the definition of modernism as the much more commonly noted tendency to "technologize aesthetics."

The aestheticist impulse in modernism continually returns to romantic notions of the aesthetic—or of beauty, at least—as an eternal or spiritual realm, unchanging and whole. Yet, although romantic aesthetics generally figured the wholeness of the aesthetic object in terms of organic metaphors, as having a kind of indivisible life or spirit of its own, modernist aesthetics attempts to reconcile the aesthetic with the technological. To this end, it often connects the spiritual and the technological, attempting to impart a sense of wholeness and the eternal to technological forms. Thus, mathematical and abstract geometric forms are figured as having spiritual attributes, as reflecting eternal forms and values. Often, as in Bruno Taut's Glass Pavilion, these aestheticized technological forms were explicitly designed as a kind of spiritual edifice, a symbol of unity for the fragmented modern city. Through this aestheticized technology, not only is the aura of the artwork maintained, but there is often an attempt to extend it to society in general, as a means of reinvesting modern society with a sense of spirituality and wholeness.

Modernism, however, never seems able to recognize the shift in the conception of technology that begins in its own attempts to merge technology and art. It continues to conceptualize technology almost entirely in the terms of instrumentality and functionality. The modernist desire to "technologize art" is, in fact, based on its desire to make art practical, functional. Engineering and mass production come to be seen as models for a new artistic production, which would turn away from bourgeois aestheticism in favor of a more technological and supposedly more democratic approach. Mass production, in other words, tends to be equated with the good of "the masses." In this approach, a house is to be a mass-produced "machine for living in"[12] and the object of design is to be "of no discernible 'style' but simply a product of an industrial order like a car, an aeroplane and such like."[13] Thus, at the level of production, art is to be subjected to a standardization and rationalization similar to that of the Fordist factory, while at the level of use, the artistic object is increasingly conceptualized in practical or functional terms. In both cases, an instrumental or technological rationality is to be applied to art, stripping it of superfluous ornamental and ritual value. The result is a new "machine aesthetic" in which form is to follow function.

Viewed in this sense, artistic modernism can be seen as simply a continuation of the larger "project" of modernity, generally taken to begin with the Renaissance rise of a rationalist, scientific-technological conception of the world. This view of modernity, however, is based on a distinction of modernity from what is seen as an older, mythical, or magical thinking, which perceives the world as animated or "enchanted" by a spirit or essence beyond human rationality and control. Modernity has therefore presented itself as a rational "enlightenment" of a world shrouded in the darkness of myth and superstition, as a disenchantment or demythologizing that divests the world of any magical essence or spirit. In effect, modernity strives to "kill" the "spirits" that animate the world, to render the objects of the world as "dead," and therefore liable to rational use and control. In fact, it is only with the "death" of magical or animistic beliefs that the utopian project of modernity—the dream of rational enlightenment, of scientific-technological progress—can be "born."

In a very similar way, artistic modernism has been seen—and, especially in the case of the 1920s avant-gardes, has seen itself—as demythologizing or destroying the magical or ritual value of the aesthetic sphere. This "technicist" tendency is obvious in the work and statements of various avant-garde movements, from Soviet Constructivism and Productivism to de Stijl to the Bauhaus. In quite similar terms, of course, Walter Benjamin would later trace the destruction of the artwork's "aura" to the rise of technological reproducibility. Benjamin, in fact, will find the modernist "emancipation" of "constructive forms" from art directly analogous to the freeing of the sciences from philosophy in the Renaissance.[14] According to this view, then, just as modernity's scientific-technological, instrumental view of the world is predicated on the "death" of animistic, magical, or spiritualized conceptions of the world, so too is artistic modernism premised upon the "death" of the aura, which Benjamin defines precisely as that sense of an autonomous, "living" spirit that "animates" the work of art.[15]

Modernism, then, equates technological reproduction—and its related techniques of assemblage, collage, and montage—with the rationalization and functionality of mass production. Montage and assemblage techniques are seen as analogous to the practices of factory assembly lines, and their "products" are viewed as similarly functional. As Peter Wollen notes of Walter Benjamin's theories, "His modernist transformation of aesthetics is founded on the postulate of Fordism, capitalist production in its most contemporary form. Just as the Model T replaces the customized coach or car, so the copy replaces the original."[16] Viewed in these terms, the very idea of technological reproducibility and assemblage comes to be seen as inherently functional—as does any object made using such techniques.

Yet, this belief in "functional form," in a "machine aesthetic," betrays the extent to which modernism misunderstands its own "aesthetic" uses of technology. Indeed, modernist aesthetics is very often based on "the myth of functional form." Taking technology and mass production as models for art and artistic production does not, after all, make modernist art inherently more functional. As Reyner Banham has shown in discussing architectural modernism, its "functional forms" were rarely particularly technological or functional; they merely "looked" technological, functional.[17]

The analogy that modernism attempts to draw between the functionality of mass production and technological reproducibility is similarly flawed. In both cases, modernism conflates productive functionality with efficacy of use or representational efficacy. Although rationalization and standardization may make factory production, and perhaps its products, more functional, the efficacy of, for example, a photograph or film is only minimally related to the rationalization and standardization of its production.

The "machine aesthetic" of modern design was, then, precisely that: an aesthetic, a style, a simulation of the rationalized, standardized forms of machines and factories, often abstracted from any functional or instrumental context. Here, the "aesthetic" of functional, technological form leads modernism—albeit unknowingly—to a conception of "technology" that is less a matter of functionality or instrumentality than of style, of aesthetics. The machine aesthetic's simulation or reproduction of "technological style" enables technological form to be separated from function; it allows a technological style or aesthetic to be "freed" or "unsecured" from its previous, functional context. This capacity for simulation or reproduction is only enhanced by the rise, so crucial to modernist aesthetics, of technological reproducibility. If the machine aesthetic's reproduction of technological style splits style from function, with the rise of technological reproducibility, the function of technology itself begins to become a matter of reproduction, of simulation.

Yet, as modernism begins to link the aesthetic and the technological, the two begin to become confused. Even as the conception of technology begins in modernism to undergo an "aesthetic turn," so too does the conception of "the aesthetic" undergo its own "technological turn." Modernism's efforts to "kill" the aura, to make art more functional and more technological, may indeed be seen as an attempt to extend an instrumental or technological rationality to the realm of art, and to cultural forms more generally. Yet this extension itself leads to a "turning" in the notion of both technology and the aesthetic. In "aestheticizing" the functional and the technological, modernism separates technological form from function; it allows stylistic or aesthetic elements to be "unsecured" from their previous context and to be recombined or reassembled into new configurations according to the dictates of "style," of "aesthetics." Yet, the "aesthetic," as it comes to be seen in terms of the technological, moves away from romantic notions of wholeness and spiritual value; in other words, it loses its sense of aura. As such, the aesthetic will become indistinguishable from culture more generally. The aesthetic, in short, becomes a matter of style, a technological or techno-cultural style. Here, both the technological and the aesthetic have become techno-cultural.

The Aesthetics of High Tech

The high-tech aesthetic obviously draws heavily from, and in fact develops out of, the modernist "machine aesthetic." In both, technology is reproduced as an aesthetic phenomenon, as a look or style abstracted from a functional or instrumental context. The modernist machine aesthetic, however, continues—at least at an explicit level—to hold to the myth of functional form. It never acknowledges that, in its abstraction and reproduction of technological form, its "aestheticization" of technological style, form has been separated from function. In the high-tech aesthetic, on the other hand, this separation of technological form and function is often readily apparent, as in the definition of "high-tech design" as "a style or design or interior decoration that uses objects and articles normally found in factories, warehouses, restaurant kitchens, etc., or that imitates the stark functionalism of such equipment."[18] Here, as in the "machine aesthetic," it is the abstraction and reproduction—the simulation—of technological forms or elements that "turns" them into stylistic or aesthetic elements, into a high-tech style. Yet, the high-tech aesthetic is not simply a matter of the reproduction, and consequent "aestheticization," of technological forms. It involves a much more general process of technological reproducibility, in which it becomes possible for any cultural form or element to be abstracted or unsecured from its previous context—videotaped, digitized, reproduced, altered, and reassembled. As it is generalized throughout contemporary culture, this process of reproduction can no longer be seen as determined by some notion of functionality; rather, it takes on its own "aesthetic" logic, replicating, recombining, and proliferating. Shorn of both its aura and its use-value, aesthetic production becomes indistinguishable from cultural production. It becomes, in other words, a process of pastiche.

Because it is defined by this process of technological reproducibility and pastiche, the high-tech aesthetic should not be viewed as a particular style or stylistic tendency. The notion of "high-tech style" has been applied to everything from starkly minimalist, "functionalist" interior design to the complex circuitry of the microprocessor. Yet, high-tech minimalism and high-tech complexity do have more in common than their use of *high-tech* as an adjective. Minimalism and complexity may in fact be seen as the two basic, and related, aspects of high-tech style or aesthetics. The tendency of high tech toward minimalist design, inherited from aesthetic modernism, is actually an extension of modernity's tendency to technologize or instrumentalize the world, to abstract and reduce it into ever more minimal, more controllable forms. It is this process that leads to the increasing technological reproduction and digitization of the world, its reduction into increasingly smaller, and supposedly more manageable, "bits" of data.[19] Paradoxically, however, as ever more data is produced, this process inevitably leads to a multiplication of the very elements it attempts to control. This proliferation of data, then, leads to an increasing level of complexity. In precisely this way, the minimalist tendency of high-tech aesthetics is inextricably linked to the complexity that is also associated with high tech.

At a certain point, this process of ever-increasing technological complexity begins to appear as a kind of cultural mutation. It begins to seem, and perhaps to become, autonomous, beyond the ability of humanity to know and control. At this point, technology becomes techno-cultural. It is precisely this sense of an incomprehensible techno-cultural complexity that is figured in the integrated circuits and microprocessors that make up the interior of most high-tech devices, as well as in all those figurations of an immensely complex circuitry or informational matrix made popular by postmodern and cyberpunk science fiction. Indeed, the sense that a techno-cultural mutation has taken place often seems directly related to the sense of being immersed in this sort of technological complexity—to that commonly observed sense that "we are already in the matrix."

The Technological Memory

This sense that a cultural, or techno-cultural, mutation has "already" taken place often seems like—and in fact has often been—the stuff of science fiction. Yet, it is not limited to science-fiction texts; it also underlies much of "postmodern theory." But then, when theorists such as Donna Haraway speak of "a kind of science fictional move, imagining possible worlds," and science-fiction writers such as William Gibson and Bruce Sterling suggest that our world has already become science-fictional, the distinction between theoretical and science-fiction texts seems to have become less and less the point.[20] Indeed, this intermingling of "theory" and "science fiction" may itself be seen as a "mutation" that results in a more complex, hybrid or—in deference to Haraway—"cyborg" form. Yet, this mutational, "science-fictional" theory is also a response to the complexity of the techno-cultural world, which makes the traditional position of the theorist—the position of an active, knowing subject distanced from a passive object-world—more and more untenable. Faced with a complexity that seems to defy any totalizing, "theoretical" comprehension, both theorists and science-fiction writers have attempted to find new ways to theorize this techno-cultural complexity, and their relation to it.

One of the most popular means of representing this relation has been to figure the human subject as immersed in a vast and inescapably complex technological space. It is precisely this figuration that links, for example, Fredric Jameson's theorizing of contemporary "postmodern space" to the depiction of near-future urban sprawl that has become so familiar in recent science-fiction literature and cinema. Indeed, the dense, "tech-noir" pastiche that surrounds the viewer in such films as *Blade Runner, RoboCop,* and *Akira,* and in cyberpunk novels such as those of William Gibson, has much in common with Jameson's vision of the "bewildering immersion" evoked by "postmodern hyperspace" (pp. 43–44). In both cases, this space is presented as a kind of mutation. In both cases, too, it is viewed as explicitly technological, not in the sense of an older, modernist aesthetic of machinery—which is present only as an allusion, as part of a pastiche of past styles—but in the sense that it is constituted through technological reproduction: it is a space of surfaces, images, simulations, empty signifiers—a space, that is, of information, of data.

It was Gibson, of course, who gave this notion of an immense, simulated data-space what has proven to be its definitive representation in his depiction of "the matrix," a future cyberspace in which "data abstracted from the banks of every computer in the human system" is given graphic representation. As in the computers and computer networks from which Gibson drew the idea, the space of the matrix is not a physical space, nor can it be figured simply in terms of technological hardware. Although computers offer various kinds of hardware for the storage of data, from chips to hard drives to CD-ROMs and DVDs, this storage space cannot be accurately described as a hardware-space. It might, in fact, better be called a media-space, as suggested by the fact that these forms of storage are known as "storage media." The space of this data is, then, a multimedia space, constituted through simulation, through technological reproduction and reproducibility. The name generally given to this simulacral, virtual space is, of course, *memory.*

Yet, what is represented in Gibson's matrix is not simply the memory-space of computers, any more than Jameson's portrait of the spaces of the Westin Bonaventure is simply a matter of architectural space. Both can, in fact, be seen as attempts to represent the "unrepresentable" space of contemporary, postmodern techno-culture. Yet, if Jameson tends to figure this space in terms of an "overwhelming," schizophrenic pastiche of images and simulations, Gibson imagines it as a kind of technological memory: a random-access memory of cultural data and styles. In this space, technology can quite literally no longer be seen as machinery, as hardware. Rather, technology becomes increasingly a matter of technologically reproduced information: images on a videotape, scenarios of a computer game, Web sites on the Internet. This is the paradox of high-tech aesthetics: as the *form* of technology edges toward "invisibility," technology increasingly comes to be seen in the form of data or media.

At an even more general level, however, this shift in the conception of technology means that, as the cultural world around us becomes ever more liable to technological, digital reproduction, any distinction between technology and culture begins to vanish. Technology comes increasingly to be seen as a matter of cultural data, as a matter of techno-culture. Technology, in short, comes to be seen precisely in terms of that pastiche of reproductions, of cultural images and data, that make up what might be called the techno-cultural memory.

Yet if, as Gibson suggests, the data of this techno-cultural memory can be randomly accessed, it cannot be accessed in its entirety. As with a computer, the data in memory must be mediated, called up on a screen. This process of screening is necessarily partial: though data may be viewed from many angles, and in many formats, though it may be processed and reconfigured, it can never be represented as a whole. Indeed, this is, perhaps, the very definition of data. But then, memory can never be fully present. As Jacques Derrida has noted, "Memory is finite by nature. . . . A limitless memory would in any event be not memory but infinite self-presence. Memory always therefore . . . needs signs in order to recall the non-present, with which it is necessarily in relation."[21] Inasmuch as memory is always mediated through signs, images, representations, it is already technological, already a matter of screening. In this sense, total recall—or at least a whole or "global" recall—is a myth.

The screening of the techno-cultural memory should not, then, be seen as a "screen memory," in Freud's sense of the term. Despite Jameson's view that multinational capitalism is a kind of inaccessible primary scene at the base of "representations of some immense communicational and computer network" such as Gibson's matrix (p. 37), there is no "real," "original," or "whole" scene that lies hidden behind the screen. All views of the data in the techno-cultural memory are partial and contingent; other combinations are always possible. Indeed, every screening creates new juxtapositions and configurations of data, new reproductions and images. The contingency inherent in the screening of techno-culture is, therefore, an effect of the proliferation of images and reproductions in memory—an effect, in other words, of what Gibson refers to as an "unthinkable complexity." The techno-cultural memory has simply become too dense, too complex, to be thought or represented as a whole; techno-culture—and with it, technology—has instead come to be seen as an ongoing process of screening, of multimedia.

This techno-cultural screening might well be seen as a new mode of Heideggerian *Entbergung,* with the same sense of an ambiguous dynamic. Although some, like Jameson, would see this screening, like Heidegger's notion of Enframing, solely as a process of regulating and securing the techno-cultural world in instrumental, or capitalist, terms, to do so is to neglect the extent to which it is also a process or movement of unsecuring, which takes place without regard for capitalist, or even for human, ends. For although, at the level of individual cases, screening a video, playing a computer game, finding information on the Web, or making money by providing these services may be quite instrumental, when the process is viewed on a larger scale, when all the complex interactions between its elements come into play, it becomes much more difficult to conceptualize as simply a matter of an instrumental rationality or Enframing. It becomes much less a case of humans screening data for their own use than of techno-culture screening itself.

Technological Life, Technological Agency

The screening of techno-cultural memory has, in an age of high tech, begun to seem beyond human instrumentality and control; it no longer seems to function according to an instrumental rationality, but according to a much more unpredictable "techno-logic" of its own. Here, then, not only has the conception of technology undergone a mutation, but technology has itself come to be seen as a mutational process or logic. The process of screening mutates the very images and data that it reproduces. By definition, this mutational process is—as all mutation is—unsettling, aleatory, beyond human prediction or mastery.

As so-called chaos theory has shown, it is precisely when a space or system reaches a certain degree of complexity that its processes become unstable, unpredictable, chaotic, that mutation occurs. As the process of screening advances, as its reproductions are themselves reproduced, and reassembled, and bits of them reproduced again, the space of techno-cultural memory becomes ever larger and more complex. Consequently, it becomes less and less likely that the complex series of interactions, alterations, divisions, and combinations possible within that space can be foreseen, much less controlled. Even within the memory-space of the individual personal computer system, it is not always possible to foresee the problems or "bugs" that result from the interaction between various types of software. When systems are interconnected into larger networks, the possibility of "bugs" tends to increase with the complexity of the overall network. Speaking of just such a "bug" that caused the 1989 shutdown of more than half of AT&T's long-distance lines, a company technology director observed: "When you're talking about even a single system, it's difficult. But when we're talking about systems of systems, then the risks are greater. All of these stored programs are interacting with each other and that makes it hellishly difficult."[22]

Such "bugs" are, of course, still a far cry from the science-fictional computers or robots that attain sentience as a result of some random conjunction of events (short circuits, lightning bolts, a spilled soft drink, etc.). Yet, the very name "bugs" suggests that a certain agency, and indeed, a certain "life," is attributed to these technological mutations. Mutation is, of course, a biological term, associated not only with life but with the reproduction of life. It is not, therefore, surprising to find the notion of a mutant technology—of technology as a process of continual reproductive mutation—frequently figured as having a life of its own. Often, such reproductions of technological life have been represented as threatening and out of control, as monstrous, or, especially given their tendency toward mutational reproduction, as cancerous or viral. Indeed, the metaphor of the computer virus suggests both the threat and the unsettling promise associated with this kind of self-replicating, mutational technological life.

Any representation of a technology that seems to have taken on its own mutational life, or at least its own agency, is bound to seem "science-fictional." Yet, these "science-fictional" representations of technological life have also, at times, invoked the return of older—or at least other—representations that have been excluded or repressed by modern scientific-technological thought, representations in which agency is not the exclusive property of a human subject. In other words, a technological life or agency that is seen as "beyond" human control or prediction often seems to invoke a sense of those "older" supernatural or magical discourses that modernity, believing itself to have surpassed, figures as "dark," "irrational," "superstitious," and "primitive."

In an age of high tech, however, this return of the magical or the spiritual in representations of technological life no longer seems to be seen as simply monstrous or threatening. Thus, for example, movements and discourses as various as techno-paganism, "new-edge" science, cyber-shamanism, and rave culture have drawn on magical, spiritual, and metaphysical discourses to figure their own relation to a technology, to a techno-cultural space or world, that often seems to have taken on a life of its own.[23] Techno-pagans, for example, see the techno-cultural world as magical, as inhabited by unseen forces, spirits, gods. They therefore interact with technology not simply as an instrument or tool, but as something with its own autonomy or agency, which is not simply under their control. Yet, they do not then see this technology simply as dangerous, "out of control," or monstrous. Their relation to it is more a matter of interaction, cooperation, respect—of allowing that technological agency to go on in its own terms, and even to be guided by it. A similar figuration of a magical or spiritual technological agency appears in William Gibson's *Count Zero*, where sentient artificial intelligences begin to manifest themselves in the matrix in the form of Afro-Haitian loa, fully capable of controlling events and guiding their human "horses."[24] As is obvious in both of these examples, magical or spiritual representations of a technological life tend to unsettle the distinction between subject and object that underlies the universalizing conception of the modern human subject and its relation to an instrumentalized world. As is perhaps also obvious, the prominence of a matriarchal spiritual discourse in techno-paganism and of Afro-Caribbean religious discourse in Gibson's portrayal of cyberspatial loa serves to suggest that the "return" of such magical discourses in representations of technology tends to involve a return of those racial and gender differences repressed by the patriarchal, Eurocentric conception of the modern human subject.

As is evident in these examples, a mutation seems to be taking place in not only the modern conception of technology but also the conception of the human subject. If modernity has defined "the human" by its status as a subject—that is, by its presumed mastery over the world—then the growing acceptance of a notion of autonomous technological agency necessarily brings that status into question. It is for this reason that attempts to figure a new relation to technology so often draw on "premodern" models, in which human beings are defined not simply by their status as active, controlling subjects, but by their connection to and participation in a world of "other" forces and agencies. In a high-tech world, this sense of "connection"—of being immersed in the techno-cultural world that surrounds us—seems to be heightened. In such a world, the human relation to technology—and with it, human identity itself—must be imagined in new ways.

Donna Haraway's notion of the cyborg is, in fact, an attempt to represent this mutation of identity, to figure a new, hybrid, and science-fictional positionality from within a techno-cultural world or space.[25] Haraway's cyborg is not, as is often the case with the more masculinist cyborgs of Hollywood films, merely a reproduction of the same old (white male) human subject, whose sense of mastery and autonomy is now protected against incursions by an armored, technological shell or body. For Haraway, the "cyborg subject position" is not stable, but mutational; it is not homogeneous or whole, but mixed, hybrid. As such, it necessarily disorganizes the boundaries between "the human" and its others, between a "living" subject and a "dead" technology. The cyborg is less a matter of identity than of a relationality that acknowledges difference within itself, rather than simply externalizing it as a monstrous other. To be a cyborg, then, is to take part in the complex reproductive processes of techno-culture, but also, by this interaction, to generate new combinations and patterns.

A similarly complex, hybrid relation to technology and to otherness can be found, as Haraway has herself observed, in the science fiction of Octavia Butler. In Butler's work, the conception of a human subject is often put into question, whether by incorporation as alien breeding partners in "Bloodchild," by the mutations brought about by an alien virus in *Clay's Ark*, or by genetic merger with an alien species in her "Xenogenesis" trilogy.[26] For the aliens in her "Xenogenesis" series, the Oankali, the distinction between technology and life has in fact ceased to exist. Technology is not for them a "dead," external instrument, but part of them. As Haraway points out, "Their bodies themselves are immune and genetic technologies, driven to exchange, replication, dangerous intimacy across the boundaries of self and other, and the power of images."[27] The "technological" bodies of the Oankali, then, are driven by a logic not of technological but of *biotechnological* reproduction, mutation, generation. Their very essence is, in fact, hybrid and mutational, defined over millennia through continual genetic merger and exchange. Literally heterogeneous, the Oankali, like Haraway's cyborg, unsettle the traditional boundaries that have defined the privileged position of the (usually white, male) human subject. Not surprisingly, given their threatened dismantling and "incorporation" of the elements of "humanity," the Oankali appear to the human characters in Butler's story as monstrous. Yet, at the same time, they offer the promise not only of survival, but of a change or mutation that will generate new and unforeseeable possibilities, a new species.

Both Butler's Oankali and Haraway's cyborg open the possibility of seeing technology differently—as something that, like "life," not only has its own agency, but contains its own generative, reproductive possibilities. Seen in this sense, technology becomes an ongoing process of mutation, of reproducing, reassembling, generating, that functions not so much in terms of "Enframing," but in terms of its own unsettling logic, its own mutational "aesthetic." In an example that almost seems designed to illustrate this idea, William Gibson, in *Count Zero,* describes an artificial intelligence (AI) that devotes itself to the creation of exquisite shadow boxes, assembled from the detritus of techno-cultural memory. Here, a techno-logic of reproduction has indeed become a form of technē, which continually unsecures and reassembles the elements of a techno-cultural world or space in a context that can only be described as "aesthetic."

Gibson's "artistic" AI is, of course, merely a metaphor for a different conception of technology, a different relation to the techno-cultural world. Yet, although this representation of a generative, "artistic" technological agency may be figurative, this is not to say that it is simply a mystification, that it is atheoretical, or simply false. It is, after all, no more figurative than Jameson's theoretical figure of a "global cognitive mapping" that would allow a broader understanding of postmodern space. Indeed, as Jameson himself suggests, in a techno-cultural space that is too complex and chaotic to be represented as a totality, such figures are perhaps the only way to theorize our relation to the techno-cultural world around us. This kind of figurative, "science-fictional" theory is precisely what Haraway, Butler, Gibson, and the techno-pagans are engaged in.

Yet, these "science-fictional" figurations should not be understood simply as the invention of human subjects; they are at least equally the result of the generative, mutational processes of techno-culture itself. If the space of techno-culture can only be represented as the complex, mutational space of memory, its processes might themselves be figured as those of a technological, or techno-cultural, unconscious. For, indeed, doesn't the logic of techno-cultural reproduction and mutation seem to follow exactly the logic of the signifier? But then, the unconscious may itself be seen as technological, if not in the instrumental sense, then in the sense that it is an ongoing process of unsecuring, of reproducing, that breaks images and other elements free of their previous context and recombines them to generate new figures, charged with both monstrosity and promise.

The position of human beings in relation to this techno-cultural unconscious cannot, therefore, be that of the analyst (or theorist) who, standing outside this space, presumes to know or control it. It must instead be a relation of connection to, of interaction with, that which has been seen as "other," including the unsettling processes of techno-culture itself. To accept this relation is to let go of part of what it has meant to be human, to be a human subject, and to allow ourselves to change, to mutate, to become alien, cyborg, posthuman. This mutant, posthuman status is not a matter of armoring the body, adding robotic prostheses, or technologically transferring consciousness from the body; it is not, in other words, a matter of fortifying the boundaries of the subject, of securing identity as a fixed entity. It is rather a matter of unsecuring the subject, of acknowledging the relations and mutational processes that constitute it. A posthuman subject position would, in other words, acknowledge the otherness that is part of us. It would involve opening the boundaries of individual and collective identity, changing the relations that have distinguished between subject and object, self and other, us and them.

This change is itself a mutational process that cannot be rationally predicted or controlled; it can only be imagined, figured, through a techno-cultural process that is at once science-fictional and aesthetic. It is only through opening ourselves to this kind of creative process, by taking part in the complex web of relations in which we are implicated, rather than simply trying to control them, that we can hope to imagine, to bring to representation, a future that, though it may seem unpredictable and alien, will inevitably be our own.

1.
The Spirit of Utopia and the Birth of the Cinematic Machine

The historical narrative of aesthetic modernism is generally taken to have begun in the mid-nineteenth century, and the figure most often cited as its progenitor—or at least its obstetrician—is Baudelaire. In such accounts, in fact, modernism's "birth" often seems to require a doctor in attendance, for it is not an entirely "natural" process.[1] The birth of modernism involves, in other words, the reemergence of an artificial or technological element that was excluded from romantic aesthetics. Indeed, Kantian and romantic aesthetics always seemed to see the idea of a technological birth as threatening, monstrous, and any doctor connected to it as either a mad scientist or a practitioner of the black arts. Thus, the primal scene of Kantian, romantic aesthetics would be precisely this *birth of the machine,* the bringing to life of technology and technique. The repression of this scene will serve to constitute the Kantian aesthetic sphere; its "renaissance" will define aesthetic modernism.

Yet, if modernism has generally been defined by the reemergence within aesthetics of technology and technique, there is still a ghost of Kantian aesthetics in the modernist machine. Most definitions of modernism emphasize the fragmentary effects—on both space and time—of modern technology. Modernism comes to be seen in terms of its openness to the urban-technological "shocks" of the modern city, to "the 'present-ness' of the present," to that "half of art" that Baudelaire characterizes as "the transitory, the fugitive, the contingent." Such definitions, however, tend to neglect what Baudelaire calls "the other half" of art: "the eternal and the immutable." Indeed, Baudelaire himself apparently sees this aspect of art as nonmodern. It would seem, in fact, to belong to a Kantian aesthetic sphere that defines itself precisely in opposition to the fragmentation and transience of modern technological life. In this opposition, the aesthetic sphere is represented as an eternal, utopian realm, in which every "object" has been endowed with the internal purposiveness, symbolic significance, and full presence of a living thing—that is to say, with a "spirit" or "soul." The transient, contingent, and inanimate technological object must therefore be excluded from this realm, as the dead are from the living. Yet, if modernism is defined by the reemergence of the technological in the aesthetic sphere, this is a realm still haunted by a transcendent, living "spirit," by the desire for the eternal and the immutable. It is haunted, in other words, by what can only be called a *spirit of utopia.*

For both modernist and romantic aesthetics, then, the birth or coming to life of the machine is not simply the product of a rational, scientific design; it is not simply a matter of construction, of putting parts together, of engineering. Rather, such a machine is necessarily infused with a living spirit, with a soul; it is a "dead" technological object reanimated, given the status of an autonomous subject. This bringing to life of technology must obviously, then, take place as much through magical or spiritual means as through science. This sort of animation of inanimate objects is common in many myths and fairy tales, from Pygmalion to Pinocchio. The animation of technology, however, tends to be figured in the terms of a dichotomy, as either utopian or dystopian. Given the often-noted "romantic reaction" against the rationalist, scientific-technological utopianism prevalent at the time, it is hardly surprising that the figure of a living, autonomous technology would appear to romantic aesthetics as almost entirely negative, dystopian. This representation will be maintained in modernism, but alongside it there is a return of the image of a utopian, animate technology.

Whether the figuration of technology as living is represented as utopian or dystopian, however, it remains a technology "animated" by a certain "spirit." Thus, in utopian representations, technology will be "spiritualized," infused with an eternal, fully present spirit of life. In such representations, which tend to draw on a tradition of mathematical and geometric mysticism that runs from Plato and Pythagorus through hermetic philosophy to Kepler and Newton, the mathematical and formal aspects of science and technology are seen as reflecting an eternal perfection and harmony.[2] On the other hand, in dystopian representations, the coming to life of technology is presented as the product of an occult or supernatural knowledge, of a black magic. The spirit that animates this technology is demonic, ghastly; it haunts technology, takes possession of it. The "dead" technological object never becomes fully living; it remains merely a simulation, undead, a technological monster or zombie. It becomes, in other words, "uncanny."

As Freud has noted, of course, the idea of a machine coming to life is frequently a source of the "uncanny" *(unheimlich)*.[3] Freud bases his analysis of the uncanny on the ambiguity in German of *heimlich*—familiar or intimate, but also concealed or secret; indeed, "the *heimlich* art" is magic; "*heimlich* knowledge" is mystical or occult knowledge. Through this ambiguity, Freud notes, *heimlich* comes to coincide "with its opposite, *unheimlich*" (ghostly, hidden, uncanny) (p. 226). Freud therefore defines the uncanny as that sense of fear experienced upon the recurrence of "something which is familiar and old-established in the mind and which has become alienated from it only through the process of repression" (p. 241). Indeed, those instances that evoke the sense of the uncanny—the idea of the double, ghosts, the return of the dead, the evil eye, the coming to life of machines or automata, and so on—represent the return of the repressed projections of a primary narcissism that is related to a magical, "animistic conception of the universe."[4] Yet, as Freud notes, not every projection that reemerges from repression is experienced as uncanny. He is never entirely able, however, to explain what factors determine whether or not an experience will be seen as uncanny. He does offer some suggestions, though, one of which is the relation of the uncanny to repressed primal fantasies, particularly to the threat of castration.

Thus, the sense of uncanniness provoked by the birth of the machine, by the coming to life of technology, can be seen as based on a threat to the "phallus," that is, on a threat to the self's legitimation of itself as a unified "subject," to its image of itself as living, autonomous, and whole; for what the phallus attempts to symbolize is precisely the authority of a unitary, living soul or spirit over the fragmentation and contingency of the object-world.[5] With the inception of scientific-technological rationality, humanity takes up this phallus (which, in medieval Christianity, could only belong to God the Father); it assumes the mantle of Cartesian subjecthood. This position of authority can only be maintained, however, so long as technology remains a "dead" object, an instrument or means to that imaginary end, that utopia in which scientific knowledge and technological control would be fulfilled. When this utopian ideal comes into question, however, technology can no longer be subordinated to human purposes or control; it becomes an end in itself—which is to say, it comes to life. Yet, to the extent that technology's life does not have the necessary significance and internal purposiveness of a fully present soul, to the extent, in other words, that it does not mirror the desired wholeness and autonomy of the self, it will be regarded as a ghastly or uncanny life, an *other* life that threatens to control or even to supplant the true presence of life.

The coming to life of the technological other, therefore, threatens to fragment the self, to mathematize and mechanize it, to make it into an object of domination rather than a subject in control. It is against this threat that the Kantian aesthetic sphere is constituted. The realm of aesthetic beauty can therefore be seen as a narcissistic projection of the self: as an imaginary, utopian space, autonomous and eternal, in which every "object" is symbolic, full of meaning, endowed with a spirit or soul that mirrors the self's own image. The aesthetic object is not only "created" in the image of its "maker" but is also, like the god the self aspires to be, "eternal and immutable." Thus, Kantian aesthetics animates the artistic object with that ritualistic or spiritual element that Benjamin designates as the *aura.* The aura is, after all, the projection of a kind of living presence or spirit onto the aesthetic object: "To perceive the aura of an object we look at means to invest it with the ability to look at us in return."[6] The look of the aura, however, is not the look of the other, but a reflection of the same. The experience of the aura, in other words, reproduces precisely the scene of Narcissus entranced by his own reflection: "[T]he painting we look at reflects back at us that of which our eyes will never have their fill. What it contains that fulfills the original desire would be the very same stuff on which the desire continuously feeds" (p. 187). This is clearly the scene of a certain "enchantment"—an enchantment that is obviously similar to the magical or spiritual animation of the cosmos that Freud (as well as Max Horkheimer and Theodor Adorno) attributes to prescientific thought. For Benjamin, too, the enchantment of the aura or the beautiful is related to a kind of magic that "conjures up" spirits out of the past: "What prevents our delight in the beautiful from ever being satisfied is the image of the past. . . . Insofar as art aims at the beautiful and, on however modest a scale, 'reproduces' it, it conjures it up (as Faust does Helen) out of the womb of time" (ibid.). The spirit of the beautiful, in other words, is conjured by drawing an image out of the past, out of that "sequence of days" that, for Benjamin, makes up history. This conjuring frees the image of the past from its subordination to a techno-teleological conception of history and yet allows it to retain the full presence of a living subject. The image comes to be seen as "animated" by an eternal, living spirit, by what Benjamin calls the "breath of prehistory" (p. 185). Benjamin associates this "prehistoric impulse to the past," to the "archaic symbolic world of mythology," with "memory, childhood, and dream."[7]

The conjuring of the aura is therefore a kind of unconscious remembrance or *mémoire involontaire* of a past filled with a "living" presence and meaning. It is, in Platonic terms, a form of *anamnesis.* In aesthetic modernism, however, the spirit or aura of the beautiful will be challenged by another type of memory: a technological memory.

In Benjamin's estimation, Baudelaire's modernity was based on his openness to the "shocks" of the modern city. These shocks represent, for Benjamin, the impact of modern industrial and technological processes on the individual. Through these shocks, he notes, "technology has subjected the human sensorium to a complex kind of training" (p. 175). As a result of technology's effect on perception, space and time come to be seen as fragmentary and transient; they can no longer be described by Kantian categories. Perception in the form of shocks therefore transforms the individual into a kind of receptive machine, into "a *kaleidoscope* equipped with consciousness."[8] Benjamin characterizes this "mechanical" reception as habitual or distracted; it is the "polar opposite" of the concentration demanded by the experience of the aura. In aesthetic modernism, this distracted, technological reception corresponds to the technological reproduction of the artwork, and to the techniques of collage and montage that are based on it.

Thus, the conjuring up of the beautiful from the past, the unconscious remembrance of a full, eternal presence or spirit, "no longer happens in the case of technical reproduction. (The beautiful has no place in it.)" (p. 187). Rather, technological reproduction is a conscious recording, a kind of *mémoire volontaire*:

> [W]e designate as aura the associations which, at home in the *mémoire involontaire,* tend to cluster around the object of a perception. . . . The techniques based on the use of the camera and of subsequent analogous mechanical devices extend the range of the *mémoire volontaire*; by means of these devices, they make it possible for an event at any time to be permanently recorded in terms of sound and sight. (p. 186)

Yet, if the images of technological reproduction are recorded "permanently," they do not have the permanence of "the eternal and the immutable." They are rather, to continue Baudelaire's distinction, transitory, fugitive, contingent. They do not, as in the experience of the aura, conjure up the spirit of the beautiful; they do not carry the mystical significance and internal purposiveness of the "living" symbol. They are, rather, "dead" fragments that, removed from the fabric of tradition, have lost their magical animation, their "spirit." They are images that have been technologized, allegorized, and now have no *necessary* end or meaning. The recording of images in technological reproduction is therefore not, in Plato's terms, a "live" memory; it is not a matter of unconsciously evoking or recalling the magical or spiritual presence of the original. Technological reproduction, rather, involves a kind of artificial or technological memory, one whose images are mere copies, imitations, citations.[9] Emptied of inherent significance, these images become arbitrary, contingent, and can thus be consciously used, arranged, constructed. A formal or constructive principle is therefore "emancipated . . . from art, as the sciences freed themselves from philosophy in the sixteenth [century]."[10] This emergence of "constructive forms" will result in a shift away from the magical, symbolic aspect of art toward more "scientific," "functional" forms. Unlike in the scientific discourse of the Renaissance, however, these forms will not be conceived simply in linear terms, but in terms of fragmentation or "shocks": as collage and montage.

For Benjamin, as is well known, technological reproduction's release of "constructive forms" or "shocks" is best exemplified in the film. In film, "perception in the form of shocks was established as a formal principle" (p. 175). In film, in other words, technological reproduction is not, "as with literature and painting, an external condition for mass distribution," but "is inherent in the very technique of film production."[11] Moreover, Benjamin makes the shift from a magical, symbolic perception (i.e., aesthetic perception) to a "scientific" or technological perspective the very basis of the distinction between film and painting:

> How does the cameraman compare with the painter? To answer this we take recourse to an analogy with a surgical operation. The surgeon represents the polar opposite of the magician. The magician heals a sick person by the laying on of hands; the surgeon cuts into the patient's body. The magician maintains the natural distance between the patient and himself. . . . The surgeon does exactly the reverse; he greatly diminishes the distance between himself and the patient by penetrating into the patient's body. . . . In short, in contrast to the magician—who is still hidden in the medical practitioner—the surgeon at the decisive moment abstains from facing the patient man to man; rather it is through the operation that he penetrates into him.
>
> Magician and surgeon compare to painter and cameraman. The painter maintains in his work a natural distance from reality, the cameraman penetrates deeply into its web. There is a tremendous difference between the pictures they obtain. That of the painter is a total one, that of the camera-

> man consists of multiple fragments which are assembled under a new law. ("WAAMR," pp. 233–34)

In this analogy, Benjamin clearly distinguishes between Kantian aesthetics and aesthetic modernism. The basis of this distinction is technological. The "natural," magical aesthetic outlook treats its object—the patient or the aesthetic object—as a subject ("man to man"), as a whole (a "total" picture) deserving of respect ("distance"). On the other hand, the technological or scientific attitude of modernism treats its "patient" as an object to be analyzed, penetrated, fragmented, and rearranged. The film, in Benjamin's view, extends this ability to penetrate or analyze in a way that is similar to the psychoanalytic penetration of everyday life:

> For the entire spectrum of optical, and now also acoustical, perception the film has brought about a . . . deepening of apperception. It is only an obverse of this fact that behavior items shown in a movie can be analyzed much more precisely and from more points of view than those presented on paintings or on the stage. . . .
>
> By close-ups of the things around us, by focusing on hidden details of familiar objects, by exploring commonplace milieus under the ingenious guidance of the camera, the film . . . extends our comprehension of the necessities which rule our lives. . . . Even if one has a general knowledge of the way people walk, one knows nothing of a person's posture during the fractional second of a stride. The act of reaching for a lighter or a spoon is familiar routine, yet we hardly know what really goes on between hand and metal, not to mention how this fluctuates with our moods. Here the camera intervenes with the resources of its lowerings and liftings, its interruptions and isolations, its extensions and accelerations, its enlargements and reductions. The camera introduces us to unconscious optics as does psychoanalysis to unconscious impulses. (Ibid., 235–37)

Benjamin compares this reemergence of the scientific-technological in the film to the integration of scientific and technological knowledge into Renaissance art. "To demonstrate the identity of the artistic and scientific uses of photography which heretofore usually were separated," he argues, "will be one of the revolutionary functions of the film" (p. 236). Yet the example of artistic-scientific integration that Benjamin gives at this point—the filming of "a muscle of a body"—suggests that his conception of an "artistic use" of film is very different from the magical, spiritual animation of the artwork that is apotheosized in Kantian aesthetics. This notion of an analytic, "scientific" aspect of art will, in fact, be part of that strange, Frankensteinian combination of living and dead, magic and technology, aesthetics and science, that animates the birth of the cinema.

Noël Burch has analyzed this birth of the cinematic machine—and its "pregnancy"—in a provocatively but appropriately titled article, "Charles Baudelaire versus Doctor Frankenstein."[12] Curiously, however, Burch tends to align science with Baudelaire and bourgeois ideology and aesthetics with Frankenstein. For Burch, the basis of this distinction is what he calls the "'Frankensteinian' aspiration" of bourgeois ideology: the desire to "triumph over death" by reproducing the mirror of life, the "perfect illusion" of a completed, fully present representation. To the extent that Baudelaire attacks this naturalistic, illusionistic notion of representation and affirms, at one point, the idea that photography should become "a servant to the sciences," he becomes the representative of science in Burch's schema.

Although the distinction that Burch makes is a useful one, this figuration is open to criticism because it overlooks the fact that neither Frankenstein nor Baudelaire can be assigned to only one side of this distinction. The story of Frankenstein, first of all, has an explicitly scientific-technological basis; moreover, it clearly evokes a sense of the uncanny that can hardly be considered representative of bourgeois aesthetics or ideology; it is not simply a matter of the triumph of life over death, but of a more monstrous aspiration: that of death come to life. The opposition is equally problematic in the case of Baudelaire, who, as Benjamin notes, only assigns photography to the sciences in order to separate "ephemeral things, those that have a right 'to a place in the archives of our memory,'" from the aesthetic "region of the intangible, imaginative," where only that on which "man has bestowed the imprint of his soul" can be admitted.[13] This distinction within Baudelaire's thought may, in fact, be seen as analogous to the division that Burch finds in the birth of cinema. As will become clear, then, the figures of both Baudelaire and Frankenstein are implicated in the dialectic of this birth, just as they are in the birth of modernism itself.

Like Benjamin, Burch finds a precedent for an analytic, scientific-technological aspect of photography and film in "the dialectical links that were beginning to spring up between artistic and scientific practices in the Italy of the Renaissance" (p. 7). Citing Erwin Panofsky's observation of the centrality of perspective and pictorial representation to the descriptive sciences of the Renaissance,[14] he notes the important role that the scientific need for recording and analysis plays in the development of photography and film. This analytic role, however, always seemed to stand in a dialectical relationship to the desire for a magical conjuring of a fully present, living representation. As Burch notes, "only an *analysis* of human and animal (or mechanical) movement could be of interest to true scientists" (p. 9). On the other hand, the desire to reproduce movement and life was the goal of inventors who, "like the alchemists of the Middle Ages, were busily seeking the Great Secret of Representing Life" (p. 10).

This desire for a magical-alchemical representation of life is, as Burch suggests, an aesthetic impulse. Commenting on the reception of Eadweard Muybridge's photographs, he notes how the attempt "to arrest movement, to break it down into a series of still photographs in the interests of 'science,'" challenged the "representational codes of academic art" (ibid.). Indeed, he argues that the "wedge driven by Muybridge between photography and the 'naturalistic' representational codes of the 19th century" was "absolutely crucial" to the cinema, because "it may in a sense be said that the efforts of the 'great pioneers of the cinema' were to be devoted to restoring 'beauty' to photography: the 'beauty' of painting, but also of the bourgeois theatre and novel, which had been innocently stripped away by Muybridge" (p. 11). Contrary to the aesthetic desire for a whole, living representation, the scientific approach of Muybridge and E. J. Marey is clearly technological: it sees photography as an instrument for recording and analysis, as a kind of prosthesis that would, in Marey's words, "compensate for deficiencies in our senses or . . . correct their errors."[15] Marey's main interest in cinema, in fact, seems to have been the ability of slow or accelerated motion to augment human perception. For him, there was little point in simply representing phenomena that were already available to the human eye.

For Burch, however, the distinction between a magical-aesthetic and a scientific-technological aspect of photography is not limited to the prehistory of cinema. Indeed, the main thrust of Burch's essay seems to be his attempt to read the films of Louis Lumière within the scientific tradition of Muybridge and Marey. Burch bases his argument on the idea that the typical Lumière film, such as *La sortie d'usine Lumière,* was conceived and carried out as *"an experiment in the observation of reality"* (p. 15). This view is based, first of all, on the fact that the Lumière brothers thought of themselves as, and were, researchers, "men of science." As Louis Lumière himself noted, "my labours were labours of technical research. I never indulged in what they call 'mise en scène.'"[16] This lack of mise-en-scène is evident in most of the Lumière films, where there is little attempt to stage the action—in *La sortie d'usine,* in fact, the camera is concealed to avoid affecting the results of the experiment—and the camera is simply cranked until the film runs out. Moreover, Lumière's framing, because it "leaves such ample space for the action to develop in all directions," suggests "a quasi-scientific attitude":[17] as Burch notes, "His scenes seem in fact to unfold before his camera rather like a microbic organism under a microscope, or like the movement of the stars seen through an astronomer's telescope" (p. 18). The result, in Burch's view, was

> a radicalisation of the "polycentrism" of a photographic image, which was itself already exempt from the centrifugal rules of academic painting. In other words, the image in *La sortie d'usine* does not, any more than the street scenes and other "general views" which followed, offer a spontaneous key to a reading enabling one to itemise the complex content, especially after a single viewing. (P. 16)

The complexity, polycentrism, and aleatory qualities of the Lumière films can be seen as roughly equivalent to the fragmentation and contingency that Baudelaire found in the modern city and its crowds. And indeed, most of the Lumière films are set in an urban environment with crowds of people moving more or less at random through the frame. The transient, ephemeral quality of the images that are recorded in these scenes also places them on the technological side of the division that Baudelaire erects between photography and art; they are clearly not part of that aesthetic "region of the intangible" that Baudelaire characterizes as "eternal" and "immutable." These images have none of the necessary internal purpose or spiritual significance of the realm of the beautiful. Nor are they subordinated to a narrative end or whole. Their complexity and transience do not, as Burch notes, easily allow them to be totalized: "Only an exhaustive inventory of everything one sees in the image (which is actually what the first reviews attempted to provide) could answer this question [what is the content of the image?]; and the result could be a text at least as wordy as Raymond Roussel's *La Vue* for each of the Lumière films" (ibid.).

Thus, the images of the Lumière films serve as documentary observations and descriptions of the "details of familiar objects," of "commonplace milieus." In recording these details, these images can be seen to have, in Baudelaire's words, a "place in the archives of our memory"—a memory that is, however, conscious, fragmentary, and technological rather than unconscious, whole, and living.[18] In other words, these images allow us, as Benjamin says of the scientific aspect of film, to extend

> our comprehension of the necessities which rule our lives [and] . . . to assure us of an immense and unexpected field of action. Our taverns and

> our metropolitan streets, our offices and furnished rooms, our railroad stations and our factories appeared to have us locked up hopelessly. Then came the film and burst this prison-world asunder by the dynamite of the tenth of a second, so that now, in the midst of its far-flung ruins and debris, we calmly and adventurously go traveling.[19]

Like all technological reproduction, then, film is a technology that exposes the familiarity of perception and the world to "shocks"; it fragments both space ("ruins and debris") and time ("the dynamite of the tenth of a second") into images that can be cited, remotivated, rearranged. For Benjamin, this fragmentation and rearrangement—perception in the form of "shocks"—is intrinsic to film's basic structure, to its *montage.* The Lumière films, which are generally composed of one shot, might seem to stand outside this structure; they have no editing, no montage. It should be remembered, though, that montage does not only mean editing, particularly in German where the term can refer not only to successive but to simultaneous assemblage.[20] In this sense, the complexity, polycentrism, and randomness of the images of the Lumière films can be seen as a kind of montage that fragments or gives a shock to the totalizing visions of a spiritual-aesthetic representation.

Yet, the Lumière films can also be, and indeed have been, read as totalities, as representations of life that can overcome the fragmentation of death.[21] Both Benjamin and Burch are aware, in fact, that the dead fragments of a scientific montage can always be reunified into a living whole, that the spirit of beauty can be, as Burch suggests, restored to photography by the "great pioneers of the cinema." Yet this restoration of the dead to life, of the fragmentary ruins of technological reproduction to aesthetic wholeness, can only take place through magical or spiritual means. Benjamin, after citing the examples of Abel Gance comparing the film to hieroglyphs and of Armand Séverin-Mars "speaking of the film as one might speak of paintings by Fra Angelico," observes that the "desire to class the film among the 'arts'" causes a number of "theoreticians to read ritual elements into it—with a striking lack of discretion."[22] Yet the theoretician who best fits Benjamin's description here, and the one against whom Burch's arguments are obviously aimed, is André Bazin.

Bazin, in fact, links the cinema (and painting and sculpture as well) to the religion of ancient Egypt, where the desire to preserve life against death led to what he calls "a mummy complex."[23] The mummies and statuary of the ancient tombs were, he argues, intended for the magical purpose of "the preservation of life by a representation of life"; representation, in other words, became a means to an "eternal" life: "To preserve, artificially, [one's] bodily appearance is to snatch it from the flow of time, to stow it away neatly, so to speak, in the hold of life" (pp. 9–10). In Bazin's view, this desire to preserve life persists in all the plastic arts, including photography and cinema. In them, one can "discern man's primitive need to have the last word in the argument with death by means of the form that endures." For Bazin, however, this "primitive need" is translated into "a larger concept, the creation of an ideal world in the likeness of the real, with its own temporal destiny" (p. 10). It is this concept of a mythical, ideal world—in short, the concept of utopia—that has determined the development of cinema as the fulfillment of what Bazin has called "the myth of total cinema."[24]

Bazin, therefore, discounts the importance of science in the birth of the cinema: "The cinema is an idealistic phenomenon" that "owes virtually nothing to the scientific spirit" (p. 17). The "begetters" of cinema, as Bazin calls them, were "indeed more like prophets"; they imagined or dreamed or conjured the cinema as the mythical world of "a total and complete representation of reality": "There are numberless writings . . . in which inventors conjure up nothing less than a total cinema that is to provide that complete illusion of life which is still a long way away" (pp. 19–20). Indeed, Bazin explicitly notes the magical, ritual, or mythic elements of this "conjured" illusion, whose origins "must be sought in the proclivity of the mind towards magic" (p. 11).

Certainly, then, Bazin falls into Benjamin's categorization of theorists who "read ritual elements" into the cinema. Indeed, in projecting the space of a total cinema, Bazin is himself taking up the role of the magician—in contrast to the surgeon with whom Benjamin equates the cinema—who conjures up a fully present and eternal spirit of life out of mythic prehistory. It is this belief in a magical overcoming of death through a "living" representation that Burch refers to—somewhat inaccurately—as "Frankensteinian." And yet, the life that Victor Frankenstein begets can, in many ways, provide a model for the birth of cinema, and for the birth of modernism in general. In other words, the birth of cinema can be traced not so much to a "mummy complex" as to what might be called a "Frankenstein complex."

Although the "mummy complex" and the "Frankenstein complex" obviously share the sense of overcoming death, they can be distinguished on the basis of their relationship to technology. In the representations of the mummy complex, every effort is made to elide technology, to erase the distinction between the original and its reproduction. In order to achieve, in other words, a fully present, living representation, the mummy complex demands that the conception of representation as a technology be effaced, for it is only when representation becomes a magical or alchemical process that the fragmentation and contingency of the dead copy can be conjured or transmuted into a living totality; only through this repression of the technological can a representation become animated by the enchanted spirit or aura of the original.

In the Frankenstein complex, on the other hand, what comes to life is precisely technology. To be sure, this animation is not entirely scientific, for, in scientific terms, technology can only be a dead instrument, a means to an end that is external to technology itself. When, however, technology ceases to be seen as subordinated to an external purpose, Frankenstein's monster is born. Of course, unlike his later movie incarnations with their extruding bolts and mechanical movements, the monster of Mary Shelley's novel is not strictly speaking technological; he is not a robot or android. He is, however, the product—at least in part—of science; he is constructed, built, from dead fragments by a "doctor."[25] Yet, as Benjamin notes, the magician "is still hidden in the medical practitioner." And this magical or alchemical aspect is certainly apparent in the case of "Dr." Frankenstein. He, like the alchemists to whom Burch refers, seeks the "Great Secret" of life—and discovering it, keeps that knowledge secret, occult. Indeed, Frankenstein himself recalls his childhood desire to discover the "metaphysical" secrets of "heaven and earth":

> I confess that neither the structure of languages, nor the code of governments, nor the politics of various states possessed attractions for me. It was the secrets of heaven and earth that I desired to learn; and whether it was the outward substance of things or the inner spirit of nature and the mysterious soul of man that occupied me, still my inquiries were directed to the metaphysical, or in its highest sense, the physical secrets of the world.[26]

Moreover, Frankenstein notes how this desire for secret or occult knowledge was fired by his readings of the magical and alchemical works of Cornelius Agrippa, Paracelsus, and Albertus Magnus, which seemed to offer a "deeper," more complete knowledge of the secrets of life than that of mere science:

> [The scientist] had partially unveiled the face of Nature, but her immortal lineaments were still a wonder and a mystery. He might dissect, anatomize, and give names; but, not to speak of a final cause, causes in their secondary and tertiary grades were utterly unknown to him. . . .

> But here were books, and here were men who had penetrated deeper and knew more. I took their word for all that they averred, and I became their disciple. . . . Under the guidance of my new preceptors I entered with the greatest diligence into the search of the philosopher's stone and the elixir of life; but the latter soon obtained my undivided attention.[27]

Beneath the scientific pursuits of the adult Frankenstein lies a childhood fascination with the secrets of a magical or alchemical creation of life. Behind the technology employed by the medical doctor, there remains, as Benjamin suggests, the magical conjurings of a different kind of doctor, whose model, as Benjamin also suggests, is Dr. Faustus. This figure of the Faustian doctor-magician will haunt, in particular, the Expressionist cinema, from Dr. Caligari to Dr. Mabuse to James Whale's Dr. Frankenstein. These figures not only bring inanimate or technological objects to life; they also reduce humans to the level of automata, transforming them into somnambulists, zombies, or the like. In either case, however, the result is a combination of the scientific-technological and the spiritual-magical that is figured as a form of technological life, as a living machine. The birth of this living machine will therefore be attended by both the magician and the medical doctor or scientist, even if they are often condensed into a single figure such as "Dr." Frankenstein.

The Frankenstein complex, then, figures precisely this combination or condensation of the magical and the technological. In it, technology will be both animated and haunted by that ritual, magical spirit of life that Benjamin so clearly connects to aesthetics. And if this technological life is frequently, as in Expressionism, represented as dystopian, monstrous, or uncanny, it is to the extent that its life is seen as fragmentary, technological, dead. It becomes an other life that threatens to take the place of the truly living. To the extent, on the other hand, that the technological can be totalized or, as Benjamin would say, aestheticized, it takes on the spirit of the beautiful, of life. Indeed, in Shelley's novel, the sole basis of Frankenstein's repeatedly voiced aversion to his creation is its ugliness. This ugliness, in fact, seems to be linked to the fragmentary, technological nature of the monster, to the skin that "scarcely covered the work of muscles and arteries beneath," and to the "horrid contrast" between his lustrous black hair and pearly white teeth on the one hand and his watery eyes and shriveled complexion on the other (chap. 5, p. 57). Moreover, this ugliness is compounded by movement, by animation itself: "I had gazed on him while unfinished; he was ugly then, but when those muscles and joints were rendered capable of motion, it became such a thing as even Dante could not have conceived" (ibid.). For Frankenstein, then, the ugliness of his creature only becomes monstrous or uncanny when brought to life, a life that remains, however, fragmentary, technological and ugly rather than fully living, beautiful, and whole.

Frankenstein's view of his monster is therefore analogous to Baudelaire's aversion to photography: as Benjamin observes, "To Baudelaire there was something profoundly unnerving and terrifying about daguerrotypy."[28] Yet, as noted earlier, Baudelaire tolerated photography so long as it stopped short of the "intangible, imaginative" realm reserved for art "on which man has bestowed the imprint of his soul." It was only when photography presumed to take up this eternally living "soul" or "spirit" of beauty that it aroused Baudelaire's antipathy. In other words, so long as photography remained a "dead" technology devoted to recording the fragmentation and contingency of the ephemeral world, so long as it did not attempt to assume the eternal, living status, the aura, of the artwork, it could be tolerated. At the same time, however, Baudelaire, in his essay on Constantin Guys ("The Painter of Modern Life"), sets out a view of art in which the ephemeral is indeed combined with the eternal, in which the "technologies" of representation and memory are precisely what lend the fragmentary "phantoms" of the modern technological world the full spirit of life. In this essay, as Paul de Man notes, Guys "is made to serve as a kind of emblem for the poetic mind"; both he and his technique come to represent the "ideal combination of the instantaneous with a completed whole . . . a combination that would achieve a reconciliation between the impulse toward modernity and the demand of the work of art to achieve duration."[29] Thus, the figure of Guys is, in de Man's view, explicitly linked to technological reproduction: "Like the photographer or reporter of today, he has to be present at the battles and the murders of the world not to inform, but to freeze what is most transient and ephemeral into a recorded image" (p. 157). And, as Baudelaire's description of Guys's working method makes clear, this technique of representing the present *(la représentation du présent)* is also a technique of *memory*, a *mnemotechnics* that, through "extraction" and "recording," recalls the transient "phantoms" of modern things, like Lazarus, to life:

> Thus two elements are to be discerned in Monsieur G[uys]'s execution: the first, an intense effort of memory that evokes and calls back to life—a memory that says to everything, "Arise, Lazarus"; the second, a fire, an intoxication of the pencil or the brush, amounting almost to a frenzy. It is the fear of not going fast enough, of letting the phantom *(fantôme)* escape before the synthesis has been extracted from it and recorded.[30]

This raising of the dead, then, is clearly not scientific. It is, rather, a kind of magical or alchemical technology that conjures or transmutes the eternal spirit of beauty from out of the transience of modernity.[31] But art, for Baudelaire, is not only living; it is, as Matei Calinescu notes, also mechanical:

> Impatient with the way many romantics described art objects in terms of organic processes, [Baudelaire] has a clear and highly significant bias in favor of *mechanical metaphors*. . . . Separated from its utilitarian goal, a machine can become an object of aesthetic contemplation, and a work of art is not downgraded when it is compared to a machine. "There is no chance in art as there is no chance in mechanics," the author of "The Salon of 1846" remarks. And he goes on to say that "a painting is a machine whose systems are all intelligible for an experienced eye."[32]

For Baudelaire, art itself is determined by the Frankenstein complex: it becomes a living machine. Aesthetic modernism will continue to define itself in these terms—albeit not always consciously. As a living machine, then, modernist art is neither entirely technological nor completely living. It is not technological in the modern sense; it is not merely an instrument to some "utilitarian goal." Yet neither is it able to achieve the enchanted, fully living presence of the aura. Art becomes, instead, a closed "system" of mechanical or even mathematical techniques that extract and record the fragmentary "shocks" of the ephemeral world. It becomes, in other words, a system of montage, of technological reproduction: a technological memory or technological life that, from the point of view of the aura, is merely supplementary, simulacral, prosthetic. It becomes a bachelor machine.

The term "bachelor machine" *(machine célibataire)* was coined by Marcel Duchamp in reference to his *Great Glass: The Bride Stripped Bare by Her Bachelors, Even.* The term was brought into broad usage by Michel Carrouges, who applied it not only to Duchamp, but to images from the work of Kafka, Jarry, Roussel, and others.[33] The term has since been used by Deleuze and Guattari, Lyotard, de Certeau, and Arturo Schwarz, among others.[34] In all these usages, however, what is figured in the bachelor machine is an autonomous technology that no longer answers to rational, instrumental standards. As Carrouges notes, "the bachelor machine has no reason for existing in itself, as a machine governed by the physical laws of mechanics or by the social laws of utility."[35] Thus, it appears to have its own secret laws, its own life, that can only be discovered or created by magical or alchemical means. Schwarz, in fact, suggests that "the bachelor machine's birthplace" is "the bachelor-alchemist's studio, or laboratory."[36] The alchemist's solitary or bachelor status is narcissistic; his dreams of discovering the secrets of life—of an elixir of life or the philosophers' stone—are merely metaphors for his own transmutation, his own ability to achieve an *aurea apprehensio* (literally, "golden knowledge": i.e., "perfect knowledge") or godhood. As Schwarz points out,

> The material liberation of the philosopher's gold from the common metal is above all a metaphor for psychological processes concerning the liberation of man from the contradictions of life. These contradictions stem from a dualistic conception to the universe which postulates the conflicting polarity of all natural phenomena. . . .
>
> [This] implies, amongst other things, the abolition of the man-woman conflictual duality in the integrated personality of the reconstituted Gnostic Anthropos." (P. 164)

It should be clear, however, that in the bachelor machine this "integration" or elision of difference is never achieved. In *The Great Glass,* the bride remains in a separate zone from the bachelors, stripped of any spirit of totality or life, a mere mechanical "skeleton," as are the bachelors themselves. The base metal of the bachelor machine never attains the sublimated, golden form of the aura.[37] The bachelor machine, in other words, is never a complete representation of life, but merely a technological life, an automaton whose life is the projection of its creator's, the bachelor machinist's, desire. The pleasure involved in this machine can therefore, as Deleuze and Guattari observe, "rightly be called autoerotic, or rather automatic" (p. 18). The "autonomy" of the bachelor machine, its life, is always mechanical rather than fully living. The bachelor machine always remains a supplement (remembering the sense in which Rousseau uses the term), a simulacrum that "merely" re-presents the full presence of life, a kind of technological "phantom" or fantasy. In this sense, the bachelor machine describes that simulacral technology of memory and representation that Freud referred to as the psychical mechanism or apparatus. As Carrouges notes, "Governed by the mental laws of subjectivity, the bachelor machine merely adopts certain mechanical forms in order to simulate certain mechanical effects"; it is, in other words, "the semblance of machinery, of the kind seen in dreams, at the theatre, at the cinema" (p. 21).

The connection between the cinematic and the bachelor machine is, then, a matter of projection. What both presume to project—what they, like Bazin, dream of—is a magical, fully present, and completed representation.[38] It is to this end that they frequently project an image of a mechanized woman or bride who, in Duchamp's words, will serve as the "motor" of desire for the bachelor machine.[39] Duchamp also makes clear that the status of this bride—and her pleasure—is imaginary: "the sexual life . . . of the desiring bride is purely imaginary" (p. 21). This imaginary projection of a bride whose marriage is never to be consummated is precisely what allows the bachelor machine to function as a closed, indeed masturbatory, system.

In "The Myth of Total Cinema," Bazin has suggested the conjunction of the cinema with the bachelor machine by his reference—given as an example of those "writings . . . in which inventors conjure up nothing less than a total cinema"—to Villiers de l'Isle-Adam's *L'Ève future.* Villiers's curious novel, published as a volume in 1886, concerns the invention of an ideal female android by none other than Thomas Alva Edison. It also prophesies—and this is the grounds for Bazin's reference—Edison's invention of a talking cinema several years before he actually began work on his dream of a machine that would synthesize phonographic recording with filmed movement and thus provide a complete representation of life. Thus, in *L'Ève future,* the figure of Edison becomes the bachelor machinist par excellence. As such, he, like Dr. Frankenstein, will be subject to a double inscription, inscribed, that is, between the poles of the scientific-technological and the spiritual-magical.

Edison the inventor is, as Villiers repeatedly notes, an engineer and a scientist. Yet he is also "the Wizard of Menlo Park," "the Magician of the Century," or simply "the Magician." Indeed, in his "Advice to the Reader," Villiers explicitly compares Edison to Dr. Faust. In doing so, however, he suggests that his interest in Edison is solely at the level of legend or myth, the level appropriate to "the work of Art-metaphysics that I have conceived." Thus, he concludes, "the hero of this book is above all 'The Sorceror of Menlo Park,' and so forth—and not the engineer, Mr. Edison, our contemporary."[40] This remark is, however, somewhat disingenuous, for what Villiers is after is not simply the obvious connection between the magician or alchemist and the artist-metaphysician.[41] Like his friend Baudelaire, he also sees a connection between the artist's and the scientist-inventor's ability to *technologically* represent the "ideal." In introducing the character of Edison, Villiers figures the "condensation" of these two perspectives in a machine that is itself designed to reproduce a more complete representation of life, the stereoscope:

> Edison is forty-two years old. A few years ago his features recalled in a striking manner those of a famous Frenchman, Gustave Doré. It was very nearly the face of an artist translated into the features of a scientist. The

> same natural talents, differently applied, mysterious twins. At what age did they completely resemble one another? Perhaps never. Their two photographs of that earlier time, blended in the stereoscope, would evoke an intellectual impression such as only certain figures of a higher species ever fully realize, and then only in a few occasional images, stamped as on coins and scattered through Humanity.[42]

What is at stake here, then, as Edison's "reveries" in the first few chapters make clear, is a technological *redemption.* Edison dreams of recording the sights and sounds of history: scenes from the Bible and classical mythology, portraits of famous personages and beautiful women. Nor does his "album" stop with the merely temporal: "And of course we'd have all the gods as well, and all the goddesses, down to and including the Goddess Reason, without neglecting Mr. Supreme Being! Life-size, of course!" (p. 23). And indeed, it is precisely the *full* representation of life that is in question here, a total, utopian representation that would "be able to recover, either by electricity or by some more subtle means, the undying interstellar reverberations of everything that has occurred on earth" (p. 22). Villiers's Edison envisions, in other words, a technological representation that would be fully present, a technological memory in which all of history would be brought to life or, in Benjamin's terms, "redeemed."[43]

Such a redemptive technology or machine is, of course, only a projection. Indeed, as Raymond Bellour has observed in his astute analysis of *L'Ève future,* it is the projection of the machine as God;[44] for, in Edison's view (that is, the Edison projected in Villiers's novel), "the living idea of God" is a matter of a "vital reflexive spirituality," that is, it exists only to the extent that the individual is able to project it.[45] As Bellour notes, "the machine becomes God only in and for the man who conceives it, insofar as he recognizes himself as God through the machine. . . . only the utopia invested in the machine can enable it to attain the infinite of which God is the psychic model" (p. 113). The narcissism of this imaginary projection is readily apparent, and perhaps appropriate for the man who, as Villiers puts it, "has taken Echo captive."

This same narcissistic projection will endow the female android with life. Indeed, Edison designs the android precisely as a substitute for the beautiful young woman with whom his friend Lord Ewald has fallen in love. The need for such a substitute arises from the fact that this woman, Alicia Clary, though physically beautiful, is afflicted with a "bourgeois" soul. Lord Ewald calls her a "bourgeois Goddess" (p. 36). Comparing her to the statue of Venus, Lord Ewald summarizes the problem this way:

> The sole misfortune that has befallen Miss Alicia is reason! If she were deprived of all reason, I could understand her. The marble *Venus*, in fact, *has nothing to do with reason*. The goddess is veiled in stone and silence. From her appearance comes this word: "I am Beauty, complete and alone. I speak only through the spirit of him who looks at me. . . . For him who reflects me, I am the deeper character he assigns me."
>
> The meaning of the statue, which *Venus Victorious* expresses with her contours, Miss Alicia Clary, standing on the sand beside the ocean, might inspire as her model—if she kept her mouth shut and closed her eyes. (P. 41; translation altered slightly)

Like Beauty, then, woman is supposed to be a reflective surface for the man whom she mirrors. But Miss Alicia's bourgeois reason prevents Lord Ewald from projecting onto her soul the desires that her beauty inspires. Hence his wish to remove the soul from her body—a wish that Edison will fulfill with the invention of the female android, Hadaly. Hadaly, then, will be a kind of "phantom," identical to Alicia, but "*without the consciousness with which she seemed to be afflicted*" (p. 85). She will be an "ideal" screen—not entirely blank, because Edison will place the words of "the greatest poets, the most subtle metaphysicians, the most profound novelists" in her "program" (p. 131)—onto which Lord Ewald will project his own soul; or, as Bellour has it, "an echo chamber" (p. 119). Thus, just as in Edison's speculations about God, it is a machine that allows the narcissistic circuit to be fulfilled.[46] Indeed, Edison, in an entirely Faustian speech, promises "to raise from the clay of Human Science as it now exists, a Being *made in our image*, and who, accordingly, will be to us WHAT WE ARE TO GOD" (p. 64).

One of the more remarkable facets of Villiers's novel, in fact, is the way in which a materialist and analytic scientific discourse is adduced to the service of a magical, idealist creation of an "almost eternal" life. Lord Ewald, in fact, after his first meeting with Hadaly, suspects that "I've come into the world of Flamel, Paracelsus, or Raymond Lull, the magicians and alchemists of the Middle Ages" (p. 62). Edison, for his part, promises to demonstrate "with mathematical certainty," "*point by point* and *in advance*,"

> how, making use of modern science, I can capture the grace of [Alicia's] gesture, the fullness of her body, the fragrance of her flesh, the resonance of her voice, the turn of her waist, the light of her eyes, the quality of her movements and gestures, the individuality of her glance, all her traits and characteristics, down to the shadow she casts on the ground—her complete identity, in a word. (P. 63)

This catalog of fragmented parts, however, is only a preface to the detailed description that Edison will offer of the workings of his female android. As this description continues, it becomes, as Bellour points out, "increasingly precise and improbable":

> It literally exhausts, point by point, the mechanical body of the Android while prefiguring the incarnate body of the Future Eve. In the course of this dissection, the cold precision of science is balanced by the seduction of its metaphors. Villiers perpetuates here the very old tradition of *blasons* or emblems describing the woman's body. . . . By dismembering, and by naming the parts, he makes of the female body an experimental field, an object exposed to generalized sadism. This corrosive materialism is, however, marked by an increased idealism. Woman . . . is blown up to proportions hitherto unthinkable. (P. 118)

Here, as elsewhere, the fragmentation and analysis of the female body is performed in the name of a science that seeks, through knowledge, to subdue the threat that it poses to the wholeness and life of the unified (male) self—a threat that in fact spurs Edison's work. This requires, however, that the female body be "reassembled," synthesized, in a nonthreatening form: as spirit or the ideal. To move, however, from analytic fragmentation to a completed whole, to the ideal Hadaly—whose name, Edison says, is Persian for the ideal—will require a kind of magic, a "secret" knowledge. This is the "secret" (of representing life) that Edison withholds from Ewald until the very end, but it is also a secret to Edison himself. This secret is that Hadaly is "imbued with two wills" (p. 12). One of these is Edison's—the scientific, "programmed" part of the android; the other is the contribution of Sowana—the female spirit whose soul will animate Hadaly. And Sowana is that part of Hadaly that Edison "does not know" (p. 211). In fact, he does not know her to such an extent that he remains unaware of the secret that passes between Hadaly-Sowana and Lord Ewald. This secret, which perhaps not coincidentally (as Edison seems to believe) escapes Edison's scientific surveillance, is Lord Ewald's agreement not to invoke the "other" women in Hadaly's "programming," but to love only the incarnation of Sowana's soul.[47] The scientific-technological aspect of the android, in other words, becomes merely a transparent instrument that allows the circuit of desire between Lord Ewald and Sowana to be completed: a kind of electrical conductor—like the magnetic rings that Edison uses to communicate with Sowana—that allows Ewald to project the desire that Sowana will reflect. An electrical philosophers' stone, then, that enables the Artist/Adept to achieve the status of God.[48]

This, then, is the kind of magical, alchemical conjuring that brings about the birth of the spiritual machine. Nor is it much different from the magic that animates Bazin's Faustian dream of a cinematic machine that would provide a totally present representation of life. In each case, the bringing to life of technology requires that the threat posed by an autonomous, other body, whether technological or female, be subjected to a fragmenting scientific analysis that will allow it to be sublimated, made to serve as the transparent instrument—like the stereoscope—of a living utopian whole.[49] Yet, in linking this whole to "the eternal female" (the title of one of Villiers's chapters), Villiers draws attention to its imaginary status; for the eternal feminine spirit—that is, Sowana—is defined precisely by its capacity to reflect and to stimulate the narcissistic projections of male desire. Like the bride of Duchamp's painting, Sowana is both the motor and the screen of a desiring bachelor machine, but unlike the bride, she is not technological but spiritual. She needs Hadaly's technological body to serve as the transparent instrument that conducts that desire, that focuses it. Together, then, Sowana and Hadaly will form the perfect machine for the total representation of life—a cinematic machine: motor, lens, and screen. Ironically, cinema history records that the first attempts to achieve this kind of representation were performed by yet another female machine attributed to Edison: the Black Maria.[50]

With its black tarpaper body designed to rotate to the light of the sun, the Black Maria is a machine intended both to make itself transparent and to highlight the object of the camera's gaze, to make its image unitary. Unlike the Lumière films, the images of the Edison Black Maria films—made for kinetoscopes rather than projection—are centered and easily readable: generally short sketches, stage acts, or celebrities set against the dark background of the Black Maria's walls. There is no sense of that "quasi-scientific attitude" of observation that Burch finds in the Lumière films, an attitude that can, at least, open an almost infinite fragmentation and complexity within familiar scenes. In the Black Maria films, this polycentric surplus has been banished; it is not so much a question of observation or analysis, but of staging the image, of presenting it as a whole. And if these images have not yet achieved the living presence or aura of art, if in the Black Maria the cinematic machine has not yet been fully born, the Black Maria is at least the technological embryo of a fully present, fully living cinematic machine.[51]

For the cinema to achieve this fully living presence, it must attain the wholeness and aura of art. In this dream of a living, spiritualized cinematic machine, the cinema's technological status must be aestheticized, spiritualized, in order to reflect the projections of a self that wishes to see itself as a fully present, living whole. To the extent that the technology of the cinema—its status as technologically reproduced, its basis in the techniques of montage—interferes with this narcissistic circuit, it comes to be seen as a threat to the living cinema's wholeness. Thus, the fragmentary montage that is the basis of the cinema for Benjamin, the polycentrism and complexity that Burch finds in the Lumière films, would be seen as dystopian, uncanny. The complexity and fragmentation of such images cannot be comprehended aesthetically, as a whole. They represent an aspect of technology that seems to be beyond human control, that in fact seems to threaten humanity's control. Thus, aesthetic modernism, despite its awareness of modern technological reproducibility and complexity, remains haunted by the spirit of utopia, by the desire for wholeness.

This tension between the fragmentation of "modern" technology and the wholeness of an "eternal" spirit is very much apparent in both Bazin and Villiers, as it was in Baudelaire. They tend to see the technological as dystopian, as a threat to a fully present representation. For Bazin, this threat is figured in his hostility to montage, which can be tolerated only to the extent that it does not threaten the ontological principle of cinema—that is, to the extent that it allows an imaginary projection of a fully present reality to take place. Cinema must, in other words, allow "what is imaginary to include what is real and at the same time to substitute for it."[52] Bazin's "realism," then, is of a very idealistic sort: "if what we see depicted had been really the truth, successfully created in front of the camera, the film would cease to exist because it would cease, by the same token, to be a myth" (p. 47). For Bazin, this mythic, imaginary "reality" is the essence of cinema, and as his discussions of neorealism make clear, it is a "reality" that in its purest form is based on a fully present "wholeness":

> Neorealism is a description of reality conceived as a whole by a consciousness disposed to see things as a whole. . . . neorealism by definition rejects analysis. . . .
>
> Neorealism . . . is always reality as it is visible through an artist, as refracted by his consciousness—but by his consciousness as a whole and not by his reason alone or his emotions or his beliefs—and reassembled from its distinguishable elements.[53]

Much like Villiers, Bazin attempts to reassemble the fragmentary elements of the modern world into a spiritual totality that is linked to the artist's vision or consciousness.[54] It is not difficult to see that from this point of view the techniques of montage would be viewed with suspicion;

Bazin in fact sees them as artificial, as mere "tricks" that are little different than the special effects of Méliès or trompe-l'oeil in painting.[55] Indeed, as Bazin makes clear, the artifice of montage is the artifice of the simulacrum: "the important thing is not whether the trick can be spotted but whether or not trickery is used, just as the beauty of a copy is no substitute for the authenticity of a Vermeer" (p. 46). Thus, against the simulation of the techniques of montage and technological reproduction, Bazin attempts to pose a true conjuring that would transform the dead fragments of this technological trickery into the living aura of a mythical total representation. Again like Villiers, he dreams of a spiritualized simulacrum, of a fully living cinematic machine.

Villiers also tends to see the fragmentation and contingency of the modern technological world as dystopian, as part of "a false, mediocre, and everchanging Reality" (p. 164). For Villiers, this modern reality—artificial, dead, technological—is readily apparent in the "living nullity" of the glance of the modern woman; this deadening glance, as Edison notes, is very much connected to technological reproduction: "This glance can be photographed. After all, isn't it just a photograph itself?" (p. 160). Benjamin observes a similar sense of the inhuman, deadly, or uncanny in photography: "What was inevitably felt to be inhuman, one might even say deadly, in daguerreotypy was the (prolonged) looking into the camera, since the camera records our likeness without returning our gaze."[56] In his efforts to defuse the threat of this technological and fragmentary "other life"—a deadly, uncanny life that does not reflect our gaze—Villiers employs a double strategy, both scientific and magical. He actually extends the deadening scientific-technological fragmentation of the world, of the female body, in order to magically transform it, to spiritualize it, to bring it to life. This, then, is the utopian side of the Frankenstein complex, of the bachelor machine. For Villiers, it manifests itself in the dream, so similar to that of Bazin, of a "factory for the production of Ideals" (p. 147).

For both Bazin and Villiers, this utopian dream of a magical factory, of a living, spiritualized machine, is an attempt to transform the dystopian, uncanny processes of modern technology, of technological reproduction and montage. Yet this dream can only occur by means of precisely those processes—by means, that is, of a process of substitution. Bazin, for example, recognizes that the cinema's representation can never be fully present; it depends on a substitute, a copy: "the screen reflects the ebb and flow of our imagination which feeds on a reality for which it plans to substitute" (p. 48). It is by making this substitution magical that he hopes to conjure an imaginary presence, a presence that can return the narcissistic gaze of the self:

> It is false to say that the screen is incapable of putting us "in the presence of" the actor. It does so in the same way as a mirror—one must agree that the mirror relays the presence of the person reflected in it—but it is a mirror with a delayed reflection, the tin foil of which retains the image.[57]

A delayed reflection, a deferred presence—yet again, the impossibility of a fully present representation, of achieving utopia, becomes apparent. It is even more apparent in *L'Ève future,* where the spiritualization of the technological body occurs by means of the substitution of a technological body for the living female body. Thus, as Bellour has noted, Villiers's attempts to spiritualize the deadly, artificial "nullity" of the (female) simulacrum depend on a substitution that is analogous to the technological reproduction of the cinematic machine and of modernism itself:

> The actual process of substituting a simulacrum [the android] for a living being directly replicates the camera's power to reproduce automatically the reality it confronts. Every *mise en scène* of the simulacrum thus refers intrinsically to the fundamental properties of the cinematic apparatus. Villiers has a deep feeling for this. He understood, as had Baudelaire before him (his praise of makeup, of the fake, for example) that in the age of mechanical reproduction the artificial has become a determining condition for modernity. (P. 131)

The bringing to life of the cinematic machine therefore comes to serve as an emblem of aesthetic modernism in general. On the one hand, it is figured as an uncanny, dystopian technological life, a life that remains, like Frankenstein's monster, fragmentary, artificial, ugly, a mere substitute for the fully living. On the other hand, it is figured as the spiritualization of technology, its transformation into the mirror of an eternal, fully present, fully living self—on the one hand, a fragmentary, technological phantom of life; on the other, the wholeness of a technology become living spirit. In Baudelaire's terms, then, aesthetic modernism lies between "the transitory, the fugitive, the contingent" and "the eternal and the immutable." And it is just this dialectic—the dialectic of the bachelor machine, of the Frankenstein complex—that will both animate and haunt the cinematic machine, as it will modernism in general.

2.
The Mediation of Technology and Gender

If, as many have claimed, aesthetic modernism can be defined by its relation to technology, perhaps no other single work condenses so many aspects of this relationship as does Fritz Lang's *Metropolis* (1926). There, modernism's fears of and fascination with technology are given overdetermined representation. In fact, Andreas Huyssen, in one of the more perceptive analyses of *Metropolis,* has argued that the film is an attempt to resolve "two diametrically opposed views of technology": an "expressionist view" that emphasizes "technology's oppressive and destructive potential" and the "unbridled confidence in technical progress and social engineering" of "the technology cult of the *Neue Sachlichkeit.*"[1] Indeed, Huyssen has, in another essay, suggested that the whole of the twentieth-century avant-garde may be defined by its experience—positive or negative—of technology.[2] Yet, viewing *Metropolis,* much less modernism in general, in terms of a dialectic between utopian and dystopian views of technology may obscure as much as it clarifies. It makes little sense, for example, to categorize Italian Futurists, Soviet Constructivists, and architects of the German Werkbund simply in terms of their technological utopianism. Although these movements do tend to see technology positively, their views of technology—of what technology is—are obviously different. Similarly, not all views of modern technology as dystopian define technology in the same way. Thus, rather than seeing the two conceptions of technology that Huyssen identifies in *Metropolis* as utopian and dystopian, it may be more helpful to distinguish between a rationalist, functionalist notion of technology and a notion that emphasizes the irrational, chaotic, and even the destructive aspects of technology, that sees it as a dynamic, shocking, almost libidinal force. Both of these conceptions of technology may be seen as either utopian or dystopian. If the technological utopianism of the *Neue Sachlichkeit* depends on a functionalist conception, the Italian Futurists embrace a technology that they see as dynamic and irrational. In *Metropolis,* on the other hand, both of these "technological aesthetics" are presented, contrary to Huyssen's view, as dystopian. Indeed, it is only to the extent that these two dystopian notions of technology can be "mediated," synthesized, that a utopian technology becomes possible. The desire for this mediation serves to structure not only *Metropolis,* but much of modernism. It is also the basis for the often-noted connections between *Metropolis* and Nazism.

In *Metropolis,* this mediation can be seen in the triadic structure that dominates the film, most prominently in the "head, heart, and hands" metaphor that structures the film, as it did Thea von Harbou's novel, from which the film's scenario was adapted.[3] As the "head" or "Brain of Metropolis," Joh Fredersen is presented as a figure of almost superhuman (or inhuman) rationality and efficiency. As the architect of Metropolis, he designs and constructs his "utopian" technological city along strictly rational and functionalist lines. Built into the city's utilitarian design, however, is a similarly Benthamite, and indeed panoptical, system of control. More precisely, this system, with its hierarchical structure, is closely related to the systems of "scientific management" devised by Frederick W. Taylor and put into practice by Henry Ford.[4] As in Taylorist and Fordist organizational charts, this structure can itself be represented as a triangle, with Fredersen at its "head."

In this system, workers must adapt themselves to a functional, technological rationality; they must function, in other words, like machines, in lockstep and geometric formation, their individual identities lost. Thus, the "hands" of Metropolis become, like Rotwang's prosthetic hand, mechanical, replaceable. What is therefore figured by this mechanization is an alienation of the social or political body: the hands are "cut off" from the utopian plans of the brain.[5] This dismembered or fragmented social body is seen as inorganic, technological, dead. The workers appear robotic, zombie-like, lacking the spirit and emotions that define human life. Conversely, Fredersen himself is shown as rigid and mechanical, as lacking a humane spirit, human emotions (e.g., the firing of his secretary). The "reanimation" of this body will therefore require a "mediator" that will bring these alienated, mechanical parts together, that will restore the spirit of life, make them an organic whole. This mediating third term, necessary to bring the brain and the hands together, will itself be a symbol of wholeness and spirit—the heart.[6] Thus, only Freder, as the mediating heart between his father and the workers, can succeed in rejoining the brain and hands.

Overlaid on the triad of "head, heart, and hands" is the all too obvious Christian symbology of the Son as the intercessor between God the Father and humanity. Indeed, Freder is quite explicitly presented as a Christ figure: he descends to the workers' level and takes the place of an exhausted worker (Georg, No. 11811), where he suffers and is "crucified" on the control dial of the "Pater Noster" machine, crying out to his father for relief. Yet Freder's role as the son/savior also involves him in another Christian symbolic triad, one that includes not only his father, Joh (Jehovah) Fredersen, but the virginal mother figure of Maria (Mary). Here, it would seem, Maria stands as the representative of the positive aspects of the workers/humanity, just as the False Maria seems intended to embody their negative side. Yet, between the symbolic triangle of Father/Son/Humanity and that of Jehovah/Christ/Mary, there is a certain slippage, a set of displacements that will continue to disrupt the symbolic structure of *Metropolis.*

The first displacement is that however much Maria represents the workers, it is difficult to align her with the "hands." Indeed, in contrast to the sensuality of her evil double, there is little about Maria that can be associated with the physical. Her role, in fact, seems more like that which Goethe, at the close of *Faust II,* ascribes to the "Holy Virgin, Mother, Queen": "The Eternal-Feminine [that] draws us upwards."[7] Thus apotheosized, Maria comes to take a place similar to that in the Christian Trinity of the Holy Spirit.

In terms of the "head, heart, and hands" typology, then, Maria's spirituality makes her much more a representative of the "heart" than of the "hands." It is she, even more than Freder, who personifies the emotive and spiritual dimensions that are excluded from Fredersen's world of rationality and efficiency. Similarly, the False Maria represents a demonic spirit or nature, a "dark heart": the irrational, destructive side of the emotions, of the soul. Thus, contrary to both von Harbou's premise and the repeated assertions of commentators, the heart (or the emotive/spiritual) does not mediate between the brain and the hands. The hands, in fact, are excluded, "cut off," from the process of mediation; the workers are presented simply as "manipulable" tools, as a technology equally susceptible to the emotional machinations of the False Maria as to the rationalized mechanization of Fredersen. The concluding handshake between Fredersen and the representative of the workers occurs, therefore, after the fact, the actual mediation having taken place, instead, between brain and heart, rationality and emotion, the scientific and the magico-spiritual. The reconciliation of these opposing terms is the true ideological project of *Metropolis,* and it is to this end that Freder, rather than Maria, must play the role of the mediator.

Indeed, this attempt at reconciliation is also presented as a mediation of paternal and maternal, masculine and feminine. It is hardly surprising, therefore, that the character relations, plot, and structure of *Metropolis* are dominated by the figure of the Oedipal triangle. In *Metropolis,* however, the basic Oedipal triangle has been subjected to a certain amount of condensation and displacement; it has, first of all, been split into two separate but interconnected triangles: the triad of Fredersen/Maria/Freder and the Fredersen/Hel/Rotwang triangle, which was excised from the film by its American editors.[8] In the first, the maternal figure of Maria stands in for Freder's mother, Hel, the resemblance between the two being such that Rotwang will, in his derangement near the end of the film, mistake Maria for Hel. Yet Maria is herself doubled by the Robot Maria: it is the sight of her embrace with his father that sends Freder into traumatized illness and hallucinations. Freder's mistaken interpretation of this embrace replicates the situation of Fredersen's "theft" of Hel from Rotwang. Indeed, as Roger Dadoun has suggested, Rotwang is the counterpart of Freder; he plays the role of the "bad son," the "rebellious rival son" castrated by the father: "His severed hand is punishment for his filial curiosity and establishes a female component of his personality."[9] Freder also has a "female component," which is made clear in von Harbou's novel, where Freder's difference from his father—his "soft heart"—is explained by the fact that he "is Hel's son" (p. 157). In the film, this "femininity"—which allows him to mediate between masculine and feminine, brain and heart—is manifested in his ability to sympathize (e.g., with the secretary, Joseph) and in his highly emotional reactions, which are emphasized by Gustav Frölich's hyperkinetic acting style. Thus, Freder's femininity is presented as something that is part of him, part of an integrated (or mediated) whole; he is *both* Hel's *and* Fredersen's son: "Freder is Hel's son. Yes . . . that means he has a soft heart. But he is yours too, Joh. That means he has a skull of steel" (ibid.). Thus, Freder's ability to resolve or mediate the Oedipal triangles of *Metropolis* is premised on his status as whole; because he combines both brain and heart, masculine and feminine, he lacks nothing. On the other hand (the one that has been cut off), Rotwang's femininity is not presented as the result of a mediated wholeness but of, precisely, a lack—a lack produced by the injunction of the Law of the Father, who necessarily possesses what Rotwang lacks: the woman/mother, but also the phallus. Even Rotwang's name suggests a femininity based on lack, on an artifice designed to cover this lack;

Rot-wang means, literally, "red cheek" (cf. *Wangenrot*, rouge), but the association of *Röte* with blood can also serve to suggest Rotwang's symbolic castration (the loss of his hand/phallus), which is covered over by an artificial, mechanical substitution. Moreover, Rotwang's madness, his "evil," is attributed to the loss of Hel to the father figure Fredersen, just as Freder's perception of the loss of Maria to his father provokes illness and hallucinations. Yet, in neither case are these disruptions of the rational, phallic order based simply on a "feminine" lack; rather, the source of these disruptions is a "feminine" artifice, an irrational mechanical substitute, a technological simulacrum: the False Maria.

The black, mechanical form of Rotwang's female robot—its form is feminine even before its birth, or rebirth, in the figure of Maria—serves to replace his lost love Hel, just as his black prosthetic hand replaces his lost one. Unlike Fredersen, whose possession of the woman/mother takes place under the imprimatur of a paternal, phallic law, Rotwang must piece together a simulated, mechanical copy, onto which he will conjure the shape, and the inverted spirit, of the woman/mother. Rotwang, in other words, invests this technological replacement not only with an electrical but with an emotional/spiritual charge, thereby duplicating the structure of the fetish. The technological object—itself defined by its reproducibility, by its status as a substitute—therefore stands in for both the mother and the phallus. Indeed, as Dadoun notes, the robot Maria can be seen as representing, on the one hand, "the severed phallus of Rotwang" (p. 146) and, on the other, "the female genitals," the reaction to which "indicates horror of the female organ, and . . . horror of sexuality" in general (p. 153). The robot, in other words, is represented as a sexuality or physicality that has been "cut off" from the organic body, a "hand" severed from the brain and heart: presented as fetishistic, masturbatory, and destructive, it is like the hand of Death that, intercut with the scenes of Freder's hallucinations and the False Maria's exotic dance, both plays a phallic/thigh-bone flute and swings a castrating scythe. Thus, the False Maria represents the condensation of a sexuality and a technology whose demonic or uncanny life is the product of a fetishization in which a necessarily alienated object (both sexual and technological) takes on the status of, is substituted for, an organic whole.

A similar fetishization occurs in Freder's vision of the machine as Moloch, where the specter of an alienated technology again appears in the form of a demonic mechanical life. Yet, whereas the Robot Maria figures a technology and a sexuality that are presented as irrationally destructive and out of control, the uncanny technological life of the Moloch machine seems to be the result of too much rationality, too much control. What is fetishized here is Fredersen's overrational scientific management, his urge to dominate, to control, to use and consume. As the representation of this dominating technological rationality, the Moloch machine does indeed treat the workers as objects to be used, controlled, possessed, and, quite literally, consumed. Similarly, Fredersen treats the woman/mother as an object to be possessed, stolen; indeed, it is his obsessive, "too strong" love for Hel that is said, in von Harbou's novel, to have caused her death. Thus, within the ideological strictures of *Metropolis,* the controlling, technological rationality of the Moloch machine can only be presented in terms of masculinity, of a phallic, paternal law (it is, after all, identified—in von Harbou's novel—as the Pater Noster machine). The irrational technology of Rotwang's robot is necessarily figured, on the other (prosthetic) hand, as castrated, castrating, deadly, and feminine.

It is worth noting here the extent to which these gendered technologies correspond to the poles suggested by Siegfried Kracauer in *From Caligari to Hitler*: "alternative images of tyrannic rule and instinct-governed chaos."[10] Moreover, as Kracauer has so clearly observed, this "torturing alternative" will haunt not only *Metropolis* and the German silent cinema but German society in general. It can be seen here that Huyssen's analysis of *Metropolis* in terms of a tension between a tyrannical, *Neue Sachlichkeit* technologism and Expressionist fears of a chaotic, irrational technology is, to a large degree, an extension of Kracauer's basic thesis. Perhaps even more, however, Huyssen seems to draw on Klaus Theweleit's *Male Fantasies* (*Männerphantasien: Frauen, Fluten, Körper, Geschichte,* 1977), a work that, in exploring the relations between Freikorps psychology and German society, suggests the extent to which Kracauer's tyranny/chaos division was cast in the terms of male and female, of a patriarchal order threatened by a chaotic, engulfing feminine other.[11] Theweleit sees this otherness as a representation of the unconscious and, following Deleuze and Guattari, notes that although it is frequently figured in terms of "floods," "streams," and "bodily fluids," it is also connected to the machine: "The unconscious is a flow and a desiring-machine" (1:255). Yet, curiously, Theweleit sees the "negativation of the 'mechanical'" in *Metropolis* only in connection to Fredersen's machines, which are, along with the working "masses," relegated to subterranean (i.e., unconscious) levels (1:257); he therefore overlooks the obvious connection between the "negativized" feminine machinery of the False Maria and the floods that she provokes. Huyssen, on the other hand, has clearly recognized this condensation in the False Maria of the (Expressionist) fear of an irrational, out-of-control technology and the fears of a chaotic, engulfing female sexuality. In his reading, however, the condensation of these fears in "the machine-woman" only takes place so that, by destroying the False Maria, "technology could be purged of its threatening aspects," thus completing the transition from an Expressionist view of technology as irrational and destructive to the *Neue Sachlichkeit*'s "serene view of technology as a harbinger of social progress" (p. 81). Huyssen's view, therefore, strongly parallels Kracauer's idea that the ending of *Metropolis* involves the affirmation of a tyrannical rule over the false alternative of chaos, as well as Theweleit's conception of the triumph of a tyrannical, patriarchal will over what it sees as a chaotic, feminized mass machinery.

Underlying these similar readings, however, is an even more crucial point of agreement: for in each, the affirmation of a tyrannical, patriarchal technological order, as exemplified in *Metropolis,* is seen as analogous to, or a foreshadowing of, the Nazi rise to power. Theweleit's work, of course, is explicitly concerned with this analogy. Similarly, this analogy is presented in the very title of Kracauer's work, and is made abundantly clear by the closing of his discussion of *Metropolis* with the now famous story of how Hitler, much impressed by *Metropolis,* had wanted Lang to make Nazi films (p. 164). Huyssen, in closing his analysis of the film, echoes this sentiment by noting, "It is well-known how German fascism reconciled the hands and the brain, labor and capital" (p. 81).

The point of contention in this essay is not, however, the connection of *Metropolis,* or German cinema, or German culture more generally, to German fascism. It contends, rather, that the fascistic implications of these cultural forms lies not in the affirmation of a tyrannical, masculine order over a chaotic feminine other, nor in the privileging of a *Neue Sachlichkeit* view of technology over an Expressionist view, but in the desire for a mediation that would restore coherence to an alienated, technologized modern world split by these dystopian alternatives. In *Metropolis,* for example, it is not simply the feminine False Maria but also Fredersen's functionalist, male technology—the Moloch machine—that is presented as dystopian, as a terrifying machine-come-to-life. Indeed, in the structural logic of *Metropolis,* it is precisely the fact of their engenderment that makes these technologies dystopian; for, in that logic, the feminine and the masculine machines are each represented as a threat to the "mediated" organic wholeness of the brain, heart, and hands. Each is, in other words, defined as a fetish, as the substitution of a "severed," "dead," partial object—both sexual and technological—for a whole, "living" subject. Indeed, in this logic, gender itself comes to be seen as a fetish.

Yet, if this logic figures these gendered technologies—and gender itself—as alienated, split, or severed, it also suggests that they are two sides of the same problem: the alienation or repression of the organic wholeness of nature (both an internal, "human" nature and an external nature) that takes place in a technological modernity. This technological modernity is itself seen as split between an overrational technology that represses (or mechanizes) the natural and a return of this repressed nature in the form of a chaotic, and often supernatural, technology. This view of modernity is not, however, limited to *Metropolis.* It is, for example, particularly apparent in those films Kracauer discusses that, like *Metropolis,* concern the artificial or technological creation of life.

In his summary of Paul Wegener's first version of *Der Golem* (1914), now lost, Kracauer observes that "the legend behind the film was the medieval Jewish one in which Rabbi Loew of Prague infuses life into a Golem—a statue he had made of clay—by putting a magic sign on its heart" (p. 31). In the film, the clay statue is rediscovered in modern times and comes into the hands of an antique dealer, who succeeds in reanimating it and makes it his servant. Unfortunately, the Golem "falls in love with the daughter of his master" and, infuriated by her horrified rejection of him, becomes a "raging monster" (ibid.). The Golem's life—which, like Rotwang's creation of the Robot Maria, is associated with ancient, magical, or occult powers—is artificial, technological. And like Rotwang's robot, as well as the workers of *Metropolis,* the Golem is initially a purely mechanical instrument or tool, a "dead," technological object—without a spirit or soul. Yet, also like the workers of *Metropolis,* the Golem yearns for wholeness, for spiritual and emotional fulfillment, for love; he is therefore drawn to the daughter of the antique dealer, much as the workers are drawn to Maria. It is the frustration of this fulfillment that leads him to become, much like the False Maria and the workers in their revolt, a "raging monster," libidinal, chaotic, destructive. In both films, then, this monstrosity is presented as the result of an artificial or technological life in which love, the spirit or soul, the spiritual/emotional aspects of life, have been repressed—a repression that leads to a return of the repressed, of a monstrous, uncanny life or spirit.

As a figure of both a dead, mechanical instrumentality and a threatening, monstrous technology, the Golem has obvious similarities to both the workers and the robot of *Metropolis.* The 1916 film *Homunculus,* on the other hand, presents a character who makes clearer the connection between the instrumentalizing repression of Fredersen and the return of the repressed in the Robot Maria and the workers' revolt. The initial stages of this story are, as Kracauer notes, similar to the story of *Der Golem.* Homunculus is, in Kracauer's words, "an artificial product" who has been "generated in a retort" by a "famous scientist." Although he becomes "a man of sparkling intellect and indomitable will," he is haunted by his artificial origins. Feeling like an outcast, he yearns for love, but people are horrified by him and say, "It is Homunculus, the man without a soul, the devil's servant—a monster!" (p. 32). Resentful, he becomes a repressive dictator, even disguising himself as a worker in order to incite riots that will give him the opportunity to crush the workers. Fredersen, it should be remembered, attempts to employ a similar strategy with the False Maria. Yet if, in *Metropolis,* the False Maria's destructive, libidinal life is the uncanny other half of Fredersen's repressive rationality and will, the two are combined in the figure of Homunculus. He is both a tyrant and a chaotic, destructive force. Both of these aspects seem to be linked to his lack of human emotion and spirit, of a soul, which is reflected, as Kracauer notes, in "his inability to offer and receive love." Unlike Fredersen, however, Homunculus's lack of humanity is the result of the artificial status of his life. Like both the tyrannial Moloch machine and the anarchic False Maria, he is a technological simulacrum, a fetish that threatens to replace a fully organic life. It is worth noting, too, that although Homunculus's origins are, like those of Fredersen's Moloch machine, scientific-technological, his end, like that of the False Maria, is one that is generally associated with the supernatural or occult: she is burned as a witch, and he will be struck by a thunderbolt.

Clearly, these films, like *Metropolis,* represent a certain anxiety about modernity, about the "domination of nature" by a modern scientific-technological rationality. This technological rationality is presented as a tyrannical will whose attempts to "master and possess" nature lead to a repression of nature, to the alienation or severing of humanity from (its) nature. It is, then, the repression of this nature, whether it is figured in terms of organicism, love, emotion, spirituality, or as a human spirit or soul, that defines modernity as lacking, as "split."[12] This split, moreover, is represented as gendered, as the repression of a feminine nature by a tyrannical, masculine technological will. Thus, these films' anxiety about modernity is cast precisely in Oedipal terms: as the imposition of a Technological Law of the Father—that is, as a castration anxiety.

Yet castration cuts, as it were, two ways. On the one hand is the castrated body; on the other, the castrated genitalia. In the first case, one is left with a body without desire, a purely automatic or mechanical body—"dead," without a spirit or soul. This is the state of those somnambulists, hypnotic victims, and other zombie-like figures that recur so frequently in the German silent cinema. The Golem (prior to falling in love), Rotwang's robot (prior to its transformation into the False Maria), and the workers of *Metropolis* (prior to their revolt) can all be included in this category. As is obvious, then, the second case often seems to follow from the first: desire divorced from the body seems to yield a kind of pure libido, a monstrous, out-of-control sexuality, a madness or "instinct-governed chaos" that seems to take pleasure in destruction.[13] Thus, the mechanical body becomes monstrous; the somnambulist becomes the vamp. This relationship between the robot/somnambulist and the monster/vamp is made even clearer in Kracauer's description of *Alraune* (1928), whose title character (also played by Brigitte Helm, the actress who portrays *Metropolis*'s Maria) has been created by an artificial impregnation; for Kracauer, she is precisely a "*somnambulant vamp* with seductive and empty features, [who] ruins all those who are in love with her, and at the end destroys herself" (pp. 153–154; emphasis added).

Yet, despite the relation suggested by his own description, Kracauer continues to pose the somnambulist and the vamp on either side of his tyranny/chaos division—as unrelated, alternative modes, rather than as two sides of the same "cut." For him, a mechanical or somnambulistic life is the result of tyranny; he sees, in other words, the relationship of Caligari and Cesare, Mabuse and his minions, Nosferatu and his victims, as the relationship of the tyrant and the masses. Thus, the masses become the tools of the tyrant, obedient to his will. The vamp, on the other hand, is a representative of an "instinct-driven chaos" that is associated with a complete lack of authority, leading to wild passions, social degeneration, and violence. Along similar lines, Kracauer's reading of Fredersen as the tyrant of *Metropolis* would seem, according to his schema, to demand that Rotwang, like the False Maria and the rebellious workers, be read as representing the instinctual and chaotic. Such a reading would, however, require one to neglect the fact that Rotwang shares many more similarities with Kracauer's supernatural tyrant figures than does Fredersen: his association with magic, the loss of his love, his symbolic castration, and his madness, even the fact that Rudolf Klein-Rogge played both Rotwang and Mabuse. Upon examination, in fact, it becomes clear that Kracauer's tyrant figures are often shot through with the instinctual and the chaotic (e.g., the evil of Nosferatu, the madness of Caligari, Mabuse, and Homunculus). Indeed, the "slippage" from Fredersen's tyrannical overrationality to the deranged and chaotic tyrant figure of Rotwang may be seen as analogous to the "descent" into madness and/or the supernatural of such scientists, doctors, and men of "sparkling intellect and indomitable will" as Caligari, Mabuse, and Homunculus. In each of these cases, there is a representation of an attempt to master and possess nature—one that often involves a Frankensteinian creation of an artificial or technological life—that is presented as excessive, as tyrannical. It thus exceeds the bounds of the natural, of the human, and slips into chaos and madness, evokes the supernatural or the uncanny. Here again, Kracauer's examples make clear that in these films, contrary to his thesis, tyranny and chaos are seen as related, and their relationship is presented as that of a tyrannical repression and a chaotic return of the repressed.

These films do not, then, as Kracauer, Huyssen, and Theweleit suggest, simply privilege tyranny over chaos, masculine over feminine, a utopian view of technology over a dystopian view. Rather, it is precisely the "split" between these poles—a division "engendered" by a repressive technological modernity—that they see as dystopian. Thus, these films represent a desire for a mediation that would overcome this division, that would reconcile these dystopian "halves"—that is, a repressive masculine technology and a repressed feminine "nature"—in an organic whole. This mediation, then, is an attempt to "heal" the (castration) "wound" of modernity, to overcome—and indeed, undo—the Oedipal-technological repression that is seen to engender and divide the modern subject. And it is in *Metropolis* that the mediation of this dystopian division, of a split or alienation that is presented as both gendered and technological, is given its most precise representation: that is, as a mediation that resolves the split between the twin dystopian poles of a repressive, overrational technological Law of the Father and the irrational, uncanny, and occult feminine technology of the False Maria by reintegrating a repressed feminine nature or spirit (the heart) and a masculine rationality and will (the brain). Only through such a mediation, *Metropolis* suggests, can a "severed," technological modern life be fulfilled, restored to wholeness. This restoration occurs even at an etymological level: the lost etymological significance of metropolis as "mother city" is reintegrated into the functional, modern metropolis of the Father.[14]

Indeed, *Metropolis* suggests that this utopian "restoration project" will involve a mediation not only of dystopian technological poles (the Moloch machine and the False Maria), but also of the conflicting architectural styles of the *Neue Sachlichkeit* (the New Tower of Babel) and Expressionism (Rotwang's house, the cathedral). This conflict was in fact a topic of considerable debate in German architectural circles in the years immediately preceding the making of *Metropolis,* a fact that could hardly have been lost on Lang, who once trained as an architect. This debate, as Kenneth Frampton has observed, manifested itself in the "ideological split" within the Deutsche Werkbund between "the collective acceptance of normative form *(Typesierung),* on the one hand, and the individually asserted, expressive 'will to form' *(Kunstwollen)* on the other."[15] Central to the notion of normative types was the idea of the "tectonic object," which, as Frampton notes, "was an irreducible building element functioning as a basic unit of architectural language" (p. 114). On the other hand, the projection of an architectural "will to form" was to take place according to an atectonic rather than a tectonic principle: thus, instead of depending on standardized, basic geometric units of architectural language, the architect was to create "pure organic forms" that could be regarded "as a literature without an alphabet" (p. 98).

It is not difficult, then, to see the connection between the idea of normative form and the tectonic, *Neue Sachlichkeit* aesthetics of Fredersen's Metropolis. Indeed, one can easily imagine the words of Hermann Muthesius, one of the great champions of normative types, in the mouth of Fredersen: "Essentially, architecture tends toward the typical. The type discards the extraordinary and establishes order" (p. 114). Yet, in *Metropolis,* this tendency toward standardization and rationalized order, manifested in the abstract, tectonic forms of the city's architecture, is clearly presented as repressive. Moreover, the architectural elements thus repressed are precisely those that display a marked tendency toward expressionist, atectonic, and "organic," rather than "abstract," forms: Rotwang's house, the catacombs, the Gothic cathedral, and, in the novel, the house of Fredersen's mother. Despite stylistic differences, these structures are all clearly posed in distinction to the normative forms and glass-and-steel construction of the *Neue Sachlichkeit.* Their forms are irregular rather than normative; their curved and nonparallel lines, their non-Euclidean geometry, suggest an architecture that is not constructed, tectonic, technological, but derived from natural processes and materials (wood and stone), atectonic, organic. This "organicism" is particularly apparent in their interiors, in the shadowy, irregular concaves of Maria's cavern and Rotwang's house, which, it should be remembered, are interconnected. These dark, vaginal spaces are secret, occult; they are hidden from the technological surveillance of Fredersen's scientific management. They seem, in fact, to hide a power that has been repressed from Fredersen's functional, technologically rationalized world, a power that is figured in the connection of these structures to the spiritual, to the religious or the magical.

Only the house of Fredersen's mother, which appears only in von Harbou's novel, seems not to share these occult connotations. Indeed, its organic connection to nature seems entirely simple and familiar: "a farmhouse, one-storied, thatch-roofed, overshadowed by a walnut tree . . . [with] a garden full of lilies and hollyhocks, full of sweet peas and poppies and nasturtiums."[16] Yet, as an almost perfect example of that style of architecture known as *Heimatstil,* a style whose "organic" connotations would later allow an easy match with the Nazi rhetoric of "blood and soil," this house can be connected to the other structures by the curious semantic ambiguity that surrounds those words based on the German root *Heim* (home). On the one hand, words like *Heimat* (home, native land or country) and *heimisch* (of the home, domestic, local) convey a sense of familiarity, connection, and intimacy. On the other, *heimlich* (secret, furtive) and *Heimtücke* (treachery) carry a sense of being hidden or concealed. Indeed, in his article "The 'Uncanny,'" Freud observed this kind of ambiguity within the word *heimlich* itself, which, he noted, could mean—or, at one time at least, did mean—both "familiar" or "intimate" and "concealed" or "secret."[17] The sense of *heimlich* as concealment seems in fact to have led it to be associated with notions that might seem the very opposite of home: magic, the occult, ghosts (*heimsuchen*: to haunt). Indeed, as Freud notes, *heimlich* comes to be "identical with its opposite, *unheimlich* [the uncanny]." Thus, the entire cluster of words based on *Heim* seems charged with an ambiguity similar to that which has so often been ascribed to the woman or mother.[18]

In *Metropolis,* then, the familiarity and warmth of the home of the "good" mother can be contrasted to the darkness, occult symbols, and secret passages of Rotwang's house, which also shares a number of the characteristics of *Heimatstil.* The haunted, uncanny life of Rotwang's house, with its doors that open and close of their own accord, is merely the inversion of the intimate, organic connectedness of Fredersen's mother's house, just as the False Maria is the uncanny, inverted spirit of the Good Maria. Thus, *Metropolis* demonstrates that the uncanny, as Freud's analysis indicates, is the inverted form of the familiar, or, as Freud puts it, the return of "something which is familiar and old-established in the mind and which has become alienated from it only through the process of repression" (p. 241). It is worth noting here, in fact, that *Metropolis*'s distinction between Expressionist and *Neue Sachlichkeit* architectural elements is cast precisely in terms of the "old" versus the "modern": that is, the mother's home, Rotwang's house, Maria's catacomb church, and the Gothic cathedral are all characterized as "old," as more ancient than the modern city that surrounds them. This opposition of an architectural modernity and an older tradition—linked to the magico-spiritual, the organic, and the feminine—is in accordance with Lang's claim to have imagined the film as a "conflict" between modern science and technology, on the one hand, and the occult or magical, on the other. The film suggests, moreover, that it is the repression of this older, magical element by a rationalized technological modernity that brings about—or, much like Frankenstein, brings to life—its inverted, uncanny form.

The idea of modernity repressing an older, more magical way of thinking was, however, hardly original to *Metropolis*. In fact, both Freud and Max Weber had advanced such ideas not long before *Metropolis* was made. For Weber, the "disenchantment of the world" was linked to a formal (or technological) rationality in which needs were "expressed in numerical, calculable terms."[19] This mathematization or rationalization of the world was, for Weber, fundamental both to modern science and technology and to modern capitalism. Freud also conceived the modern scientific, materialistic outlook in terms of its ability to "surmount" "the old, animistic conception of the universe," which "was characterized by the idea that the world was peopled with the spirits of human beings; by the subject's narcissistic overvaluation of his own mental processes" (p. 240). If, however, *Metropolis* accepts this notion of modernity's repression of an older, magico-spiritual conception of the world, it certainly does not affirm it. Indeed, it figures this repression in terms of loss, division, alienation—that is, as a sense of *homelessness [Heimatloskeit]*. Thus, what *Metropolis* strives for is precisely a restoration of a "familiar" sense of "enchantment" and connection to nature that strongly parallels *Heimatstil*'s emphasis on "home" and the "familiar," on the older, "*rooted* values of an agarian craft economy."[20]

Yet, the desire in *Metropolis* for a restoration of the home, of the familiar, is not simply an abandonment of modernity in favor of the ancient, the natural, the spiritual. Rather, *Metropolis* aspires to a mediation of the masculine technological will of modernity with an ancient feminine spirit or nature. Here, once again, the ending of von Harbou's novel makes explicit what is only implied in the film. In the novel, a chastened Joh Fredersen goes to the home of his mother, where he speaks of recognizing—in Freder's face—the face of Hel, of Maria, and of the masses. Thus, Fredersen's repression of the workers is linked to his oppressive love of Hel and to the paternal domination of a feminine nature. His reconciliation with this feminine nature is therefore figured as a return to *home*: as his mother tells him, "I received this letter from Hel before she died. She asked me to give it to you, when, as she said, you had found your way home to me and yourself" (p. 249).

The character of this "home" to which Fredersen returns is, perhaps, made clear by the letter itself, which speaks of Hel's "everlasting love" and ends with a quotation from the Bible: "Lo, I am with you always, even unto the end of the world." Thus, it is in an eternal, spiritualized feminine love—much like Goethe's eternal feminine—that the masculine, rational will finds its resolution, its completion, its destiny or home. In rediscovering this feminine spirit, in returning home, Fredersen (at least in von Harbou's novel) accomplishes a mediation in which his repressive, paternal rationality is spiritualized, overcome—a sublation that was already implicit in the mediated persona of his son Freder. Thus, the relegitimation of Fredersen's (and presumably, after him, Freder's) leadership at the end of the film is not simply an affirmation of a tyrannical masculine technological order; it is based on the mediation of this patriarchal order with an eternal-feminine spirit. If, moreover, one takes seriously the idea of a linkage between *Metropolis* and National Socialism, this reading of the film suggests that Hitler's appeal was also founded on a mediatory logic, on the idea of the Leader as Mediator. Certainly, Dadoun makes this connection when he notes the striking similarity of *Mittler* (Mediator) and Hitler.

It is through the leader's mediation, then, that the divisions of modernity are to be resolved, completed, made whole. The appeal of Hitler, of the Nazis, was founded on the desire for completion, for wholeness; their followers, as Theweleit notes, were

> the not-yet-fully-born, men who had always been left wanting; and where was the party that would offer them more? It was certainly not the rationalist-paternalist Communist Party....
>
> [These men] submit to orders and connect into sites that promise to eliminate what they experience as lack. The word they repeatedly scream at the party congress is "whole"—heil, heil, heil, heil, heil—and this is precisely what the party makes them. They are no longer broken; and they will remain whole into infinity. Eternal life takes place in the here-and-now... really and truly. (2:410–12)

As Theweleit makes clear, then, the wholeness that Hitler and the Nazis promise is cast in opposition to a "rationalist-paternalist" communism, which could only be associated with the alienating, mechanizing effects of a technological modernity. The Nazi wholeness, instead, involves a mediation of this repressive modernity (the here and now) with a repressed "eternal" spirit: the spirit of Germany. What the leader-mediator offers, then, is a restoration of this ancient spirit: a respiritualization, a "reenchantment," that will "reanimate" a "dead," technologized modern world. Indeed, National Socialism is cast precisely as the "rebirth" of the German spirit *(der Deutschen Renaissance)* in the here and now, a "rebirth" with obvious appeal to those whom Theweleit calls "the not-yet-fully-born."

In this scenario of rebirth, Hitler seems at times to be represented as both father and mother, as embodying both a "steely," paternal will and a spiritual-maternal emotionality and love (for Germany, for the German people). Certainly, his vision of National Socialism combines both of these attributes: "It will be unchangeable in its doctrine, hard as steel in its organization, supple and adaptable in its tactics; in its entirety, however, it will be like a religious order."[21] In this context, it is worth noting how Theodor Adorno's analysis of fascist appeal points precisely to Freud's suggestion that the unconscious "love relationships" that underlie group psychology are often expressed "through the mediation of some religious image in the love of whom the members unite and whose all-embracing love they are supposed to imitate in their attitude towards one another."[22] In Adorno's view, however, this concept of love is replaced (repressed?) in modern fascism by the "threatening authority" of the primal father; the notion of love is therefore "relegated to the abstract notion of *Germany* and seldom mentioned without the epithet of 'fanatical' through which even this love obtain[s] a ring of hostility and aggressiveness against those not encompassed by it" (p. 123). Yet, the "relegating" of the concept of love to "the abstract notion of *Germany*" would hardly seem to reduce its efficacy, particularly in view of the logic of Rudolf Hess's pronouncement that "Hitler is Germany, and Germany is Hitler." Here, yet again, a representation of Nazism in terms of a tyrannical and castrating patriarchal will neglects the extent to which the appeal of National Socialism and Hitler drew on religious-spiritual notions of love, the German spirit, *Heimat,* and so on, which were seen as linked to an "eternal feminine." Indeed, as Walter Langer has observed, "although Germans, as a whole, invariably refer to Germany as the 'Fatherland,' Hitler almost always refers to it as the 'Motherland.'"[23] Along similar lines, Anton Kaes, writing of Helma Sanders-Brahms's *Germany, Pale Mother,* notes that the film, and the Brecht poem from which it takes its title, are part of "a tradition that goes back to the late Romantic poet Heinrich Heine [in which] Germany appears as the mother."[24] In the film, Kaes argues, Sanders-Brahms attempts to distinguish between "the National Socialists of her film, the fathers who go to war: Germany as the father's land" and "the German people, Germania, the 'pale mother,' on the other hand" (p. 148). Yet it was precisely this kind of gendered division that the Nazis sought to overcome, to mediate.

This mediation, of course, is to take place through the person of the Führer, who, like Freder in *Metropolis,* combines the will of the Father (the "skull of steel") with the eternal-feminine spirit and emotions of the Mother (the "soft heart"). This conjunction may help to explain the seemingly paradoxical character of the Hitlerian persona, which ranges from an aloof, implacable sternness to the almost quivering emotionality displayed in many of his speeches. Indeed, Hitler's impassioned responses and gestures often seem to parallel Freder's hyperemotional reactions, which served in *Metropolis,* as noted earlier, to indicate the "feminine" aspect of his character. The parallels between Freder and the Führer suggest, moreover, that Hitler's mediation of paternal and maternal qualities is based less on a parental position than on that, as with Freder, of the Son. Theweleit has argued a similar position, noting that the Führer "embodies not paternal power but the common desire of sons" (2:373). Yet, as is well known, what sons desire is not simply, as Theweleit will claim, access to the father's power, but also access to that which is forbidden by the father's power: the realm of the mother. What the Führer, like Freder, promises, then, is access to both: the resolution of a gendered modernity, of sexual difference, into an *androgynous* whole—the National Socialist state. It is a resolution that takes place neither simply in the Name of the Father nor solely in the Name of the Holy Spirit, but in the Name of the Son: the Name, that is, of the Mediator.

If it is quite explicitly the Name of the Mediator *(Mittler)* that enables the resolution of *Metropolis,* it is the name Hitler that guarantees the National Socialist "(re)solution": "The party is Hitler," says Hess at the end of *Triumph of the Will,* "Hitler is Germany, and Germany is Hitler." It is, then, through Hitler that an alienated, modern Germany is to be reinfused with the eternal German spirit, mediated, made whole, transformed into a restored (i.e., National Socialist) Germany. This mediation of a technological modernity and an eternal spirit—like Freder's mediation of head and heart, masculine and feminine—is supposed to resolve the split between a repressive technological rationality or will and a repressed, technologized nature or spirit—a split that is seen as haunting not only *Metropolis,* but modernity itself.

This conception of modernity as split or alienated by a repressive technological rationality has a considerable history; it can be seen, for example, in Schiller's analysis of the modern condition, where "the wheel" of technological rationality splits "man's being," represses a part of human nature: "with the monotonous noise of the wheel he drives everlastingly in his ears, [man] never develops the harmony of his being, and instead of imprinting humanity upon his nature he becomes merely the imprint of his occupation, his science."[25] For Schiller, then, this repression of (human) nature produces a mechanical, robot-like copy or "imprint": an artificial, technologized nature. At the same time, however, Schiller sees in a technologized modern nature what can only be described as a return of the repressed: a return of those chaotic, out-of-control, and clearly libidinal "impulses" that are, for Schiller, connected to the "nature" of the working "masses": "Among the lower and more numerous classes we find crude, lawless impulses which . . . are hastening with ungovernable fury to their brutal satisfaction. . . . Society uncontrolled, instead of hastening upward into organic life, is relapsing into its original elements" (p. 35).

It is not difficult, then, to recognize in Schiller's view of a divided, technologized modern world the figures of tyranny and "instinct-governed chaos," as well as those of the robot/somnambulist and the monster/vamp. Yet an even more striking similarity, perhaps, is Schiller's desire for a mediation of these divisions, this split of modernity. For Schiller, however, this mediator—whose task, as one commentator notes, will be "to put back together the 'halves' of man that have been torn asunder"[26]—will not be the heart, but Art: "we must be at liberty to restore by means of a higher Art this wholeness in our nature which Art [i.e., artifice, technology] has destroyed" (p. 45). For Kant and Hegel, too, the aesthetic realm is defined in terms of mediation, as that which, as Hegel puts it, "stands in the *middle.*"[27] It is through art or aesthetic judgment, then, that the splitting of the modern, technologized world is to be mediated, made whole.

The aesthetic realm is, in fact, defined explicitly in opposition to the technological. In contrast to the dead, partial, technological object, which has no intrinsic meaning or purpose, the aesthetic object is seen as an organic whole, as having, like a living thing, its own, internal meaning and purpose.[28] Indeed, the aesthetic object is not presented as an object at all, but as a living subject, with its own spirit or soul, its own "aura." As

Benjamin would later note, "To perceive the aura of an object we look at means to invest it with the ability to look at us in return."[29] Thus, art itself comes to represent that old sense of magico-spiritual enchantment or animism, of an eternal spirit, repressed by the modern, technological world.

It may seem, therefore, that such a conception of art (or of the aesthetic), cast as it is in opposition to the technological and to modernity itself, could have no place in aesthetic modernism or, to the extent that it did appear, could only be represented as nostalgic or regressive. Yet, aesthetic modernism is, from its very beginnings, caught up in a desire to mediate between the eternal and the modern, the magico-spiritual and the technological, a desire, one might say, to spiritualize, or rather, to *aestheticize,* the modern and the technological. Thus, Baudelaire, despite his openness to the artifice and "shocks" of a technologized, modern nature (i.e., the city), saw the need for a mediation of this "transitory" modernity with an abstract, eternal spirit. He therefore defined art in terms of its power to "distil the eternal from the transitory," to combine two "halves"—"the ephemeral, the fugitive, the contingent" and "the eternal and the immutable"—into "a completely viable whole."[30]

This conception of art as a mediation of the modern, technological world and an abstract, eternal spirituality will appear in various guises throughout modernism. In fact, contrary to the prevailing, "functionalist" view of modernist art and architecture, modernism has frequently presented itself in religious or spiritual terms, as is explicitly the case in, for example, Neo-Plasticism, early de Stijl, Suprematism, Expressionist architecture, and the early Bauhaus.[31] Yet, in movements such as these, one can also observe, in contrast to Baudelaire, the demand—noted by Peter Bürger—that "art become practical."[32] These attempts to, as Bürger puts it, "organize a new life praxis from a basis in art" (p. 49) will therefore frequently be translated into efforts to "respiritualize" or "reunify" the modern, technologized world. If, however, it is art (or the artist) that is to reunify society, to make it (spiritually) whole, this unification will itself be cast as a mediation of art and technology.

Not surprisingly, perhaps, these demands for a "practical" aesthetic mediation of/with the modern, technological world are particularly influential in architecture. Within German architectural circles, this desire for mediation will transcend the previously noted poles of Expressionism and the *Neue Sachlichkeit,* although it will be differently inflected by each. In short, then, this difference can be stated in terms of these movements' differing views of technology: whereas the *Neue Sachlichkeit* tends to affirm an abstract aesthetic based on a rationalist, functionalist notion of technology (a technologized aesthetic), Expressionism attempts to subordinate what it sees as an irrational and divisive technology to a spiritualized, holistic aesthetic (an aestheticized technology). In the present context, however, it is the Expressionist tendency to spiritualize or aestheticize technology (and modernity in general) that is most instructive.

This tendency is exemplified in Bruno Taut's Glass Pavilion, designed for the Cologne Werkbund Exhibition of 1914. Inscribed with mystical aphorisms from Paul Scheerbart's *Glasarchitektur*, the Glass Pavilion was intended to serve as the *Kultursymbol* of a spiritual renewal in which a more practical, democratic art would be the basis for a new societal unity. Yet, Taut's utopian aspirations were, as Frampton has observed, tied to certain medieval models, particularly the Gothic cathedral:[33] "According to Taut this crystalline structure . . . had been designed in the spirit of a Gothic cathedral. It was in effect a *Stadtkrone* or "city crown", that pyramidal form postulated by Taut as the universal paradigm of all religious building, which together with the faith it would inspire was an essential urban element for the restructuring of society" (p. 116). This notion of the *Stadtkrone*, of "a new religious building capable of unifying the creative energy of the society as in the Middle Ages" (p. 118), was to prove extremely influential. It would dominate the "Utopian Correspondence" exchanged in 1919–20 by the architects of the Glass Chain, whose members included Walter Gropius. It would manifest itself not only in Mendelsohn's Einstein Tower but in Hans Poelzig's *Grosse Schauspielhaus* for Max Reinhardt. Indeed, it would even appear prominently in the Proclamation of the Weimar Bauhaus, both in Lyonel Feininger's woodcut *The Cathedral of Socialism* and in Gropius's call for a "new building of the future, which will embrace architecture and sculpture and painting in one unity and which will rise one day toward heaven from the hands of a million workers like the crystal symbol of a new faith."[34]

The unifying effect of the *Stadtkrone* was, then, based precisely on its mediatory capacity, its ability to reintegrate an "older" spirituality, associated with the Gothic cathedral and medieval society, into the here and now of a technological modernity. The glass or crystalline tower came, in fact, to be figured as a kind of spiritual radio receiver—both antenna and crystal tuner—for "tuning in with the cosmos," to use Ernst Bloch's trenchant phrase.[35] If, however, Bloch was well aware of the "incredible problems of a *'Gothic style' within the crystal*" (197), of mediating "spiritualized," organic forms and abstract, rationalist geometry, such a mediation was at the "heart" of the utopian dream of an "aestheticized" modern city, presided over by a symbolic new cathedral, the "cathedral of the future."

If this dream of a mediated, aestheticized modern city would be given what is perhaps its fullest representation in *Metropolis,* it was only in National Socialism that the idea of an aestheticized modernity would achieve its culmination: what Walter Benjamin would refer to as the "aestheticization of politics"—or more precisely, the aestheticization of the modern state. This is the dream of a spiritualized, "organic" state machinery, of a modern, technological state imbued with a soul, with an eternal spirit; it is the dream of a modern state become "a home"—familiar, whole, enchanted—the dream of a state with *an aura.* For the Nazis, in other words, the state itself was to become the cathedral of the future, in which the ancient spirit of the German people would be restored to an alienated modernity through the artist-leader's mediation, as an expression of his "will to form."[36]

This vision of the state as a work of art and of the leader as its artist had, in fact, been expressed by Joseph Goebbels as early as 1929, in a now well-known passage of his novel *Michael*:

> Art is an expression of feeling. The artist is distinguished from the non-artist by the fact that he can give expression to what he feels. In some form or another: the one in images, the other in tone, the third in words, the fourth in manner—or even in historical form. The statesman is also an artist. For him, the people represent nothing different than what the stone represents for the sculptor. The leader and the led *[Führer und Masse]* presents no more of a problem than, for instance, the painter and color.
>
> Politics is the plastic art of the state, just as painting is the plastic art of color. This is why politics without the people, or even against the people, is sheer nonsense. To form a People out of the masses, and a State out of the People, this has always been the deepest sense of a true politics.[37]

There can be no denying the extent to which the Nazis realized these monumental "artistic" ambitions, or the role played in this realization by what Goebbels would term "the creative art of modern political propaganda." The ability of this "art" to "win the heart of a people and to keep it" was, as Goebbels implicitly acknowledged, based on its mediatory capacity, on its combination of "modern," technological forms with the spirit of the German people:

> the shining flame of our enthusiasm . . . alone gives light and warmth to the creative art of modern political propaganda. Rising from the depths of

> the people, this art must always descend back to it and find its power there. Power based on guns may be a good thing; it is, however, better and more gratifying to win the heart of a people and to keep it. (From *Triumph of the Will*)

This passage, taken from Goebbels's speech at the 1934 Nuremberg rally, plays a major role in both Kracauer's explication of *Metropolis* and Leni Riefenstahl's film of that rally, *Triumph of the Will* (1935). Not surprisingly, Kracauer reads the passage, and the films, as indicating the triumph of a tyrannical will over this "heart," a reading that, in his view, is supported by the precedence that both films give to abstract, geometric forms, to the "mass ornament." Thus, he notes that the final scene of *Metropolis* "confirms the analogy between the industrialist [Fredersen] and Goebbels," and all the more, presumably, Hitler:

> If in this scene the heart really triumphed over tyrannical power, its triumph would dispose of the all-devouring decorative scheme that in the rest of *Metropolis* marks the industrialist's claim to omnipotence. Artist that he was, Lang could not possibly overlook the antagonism between the breakthrough of intrinsic human emotions and his ornamental patterns. Nevertheless, he maintains these patterns up to the very end: the workers advance in the form of a wedge-shaped, strictly symmetrical procession which points towards the industrialist standing on the portal steps of the cathedral. The whole composition denotes that the industrialist acknowledges the heart for the purpose of manipulating it; that he does not give up his power, but will expand it over a realm not yet annexed—the realm of the collective soul. (P. 164)

Thus, the crucial point in Kracauer's reading of the film, and his analogous reading of *Triumph of the Will,* is that the basis of Nazism is a subordination of "the heart," associated with "intrinsic human emotions" and "the realm of the collective soul," to a tyrannical, technological (or "industrial") will, represented by the geometric forms of the mass ornament.[38]

Despite its wide acceptance by later critics, Kracauer's reading here, as this essay has attempted to show, misunderstands the appeal of these films, and of Nazism itself. The basis of this appeal is not a "desire for tyranny"—an idea that would extend even the most pessimistic conceptions of false consciousness—but precisely a desire for a sense of wholeness, of *Heimat,* in which the alienation and divisions of a modern, technological (and indeed, capitalistic) world would be overcome, mediated, aestheticized. Thus, the "aesthetic" mediation of brain and heart, rationality and human emotions, masculine and feminine, modern technology and an ancient or eternal spirituality, is not simply "an ideological veil," hiding the "truth" of Nazi desire; this mediation is itself the "truth" of Nazi desire, of the appeal of National Socialism.[39] In other words, Goebbels and Hitler did not admire *Metropolis* because, as Huyssen suggests, they saw its notion of mediation "as a welcome ideological veil to cover up the conflict between labor and capital";[40] rather, they saw this mediation as a "true" expression of National Socialist goals and ideology.

From this perspective, *Metropolis*'s final scene does indeed represent a mediation of a tyrannical, patriarchal technological rationality with an ancient, repressed feminine spirit or nature. This mediation is evident in the very organization of the scene, which places a repentant Fredersen, his hair turned white, between the abstract, tectonic formation of workers in the foreground and the organic, atectonic form of the cathedral in the background. These aspects have, of course, already been mediated in the white-clad person of Freder, whose task it is to bring about the final joining of hands between the workers' representative and his father, a joining that, framed by the doors of the cathedral behind them, restores the feminine emotionality and spirituality represented by Maria (who stands, offscreen, to one side) to the mechanized workers (who stand, also offscreen, to the other). Through this mediatory aesthetic, then, the "mass ornament" is reconnected to its spiritual and emotional nature, to its "eternal-feminine" home.

A similar mediatory aesthetic inflects the architectonics of *Triumph of the Will*. Here again, commentators following Kracauer have tended to see the massing of crowds into abstract, geometrical patterns as indicative of a rationalization or mechanization, a triumph of a technological rationality or will. Such an interpretation neglects the fact that these crowds are not presented, as in *Metropolis*, as dispirited, soulless robots or somnambulists, but as just the opposite: a reunified, respiritualized German people, a people to whom the ancient German spirit, soul, or nature has been restored. Thus, the abstract, modernist architectonic of the mass ornament is mediated with a familiar, older spirit, a sense of organic or natural connection, of *Heimat*, which is represented by the "organic," Gothic architecture of the ancient city of Nuremberg, with its connotations of past German glories. Indeed, in *Triumph of the Will*, as Kracauer observes, "even Nuremberg's ancient stone buildings," its "steeples, sculptures, gables and venerable facades" (301–2), are mobilized in the service of the National Socialist symbology of mediation and restoration. In this "restoration," the modern, technological architectonic of the mass ornament and the ancient, organic, or atectonic forms of Nuremberg's Gothic cathedrals and buildings are mediated, reassembled into the "aestheticized" form of the Nazi cathedral of the future.

It is in this figurative cathedral that one finds the true architecture of National Socialism, the *Stadtkrone* through which the divisions of the modern world are to be mediated, the German spirit restored, the German people reunified. The architecture of this cathedral is, precisely, an architecture of mediation. It combines the forms of both the *Neue Sachlichkeit* and Expressionism; it represents itself as modern and technological, on the one hand, and, on the other hand, as the heir to a mythical Aryan tradition, of which the Gothic cathedrals are a part. But perhaps the single most accurate representation of this imaginary structure can be found in Albert Speer's "cathedral of light," a luminous "structure" formed from banners and the beams of searchlights, which can be seen in *Triumph of the Will* when Hitler addresses the assembled party officials.[41] The "architecture" of Speer's "cathedral," at once modern and ancient, technological and spiritual, can be seen as the logical extension of the premises of Taut's Glass Pavilion: a religious structure that, unbounded by material constraints, is open to both the cosmos and the people—thus allowing it to serve as a kind of aesthetic focal point, mediating between the celestial and the modern world.[42]

Yet, as Kenneth Frampton has noted, Speer's "cathedral of light" was itself subordinated to the demands of another art, an art that was, even more than architecture, central to the Nazi effort to aestheticize the modern, technological world: "Leni Riefenstahl's documentary film of the Nuremberg rally of 1934, *Triumph des Willens (The Triumph of the Will),* was the first occasion on which architecture, in the form of Speer's temporary setting, was pressed into the service of cinematic propaganda. Henceforth Speer's designs for stadia at Nuremberg were determined as much by camera angles as by architectural criteria" (p. 218). What Frampton's point demonstrates, then, is that Speer's "cathedral" was as much a cinematic as an architectural construction. Its tendency to the cinematic was not, however, implicit simply in its mise-en-scène, but also in its ability to serve as the very figure of cinema: at once a projection of light, a lens, and a screen. And is not *Triumph of the Will,* in its own way, figured as a "cathedral of light," a luminous aesthetic structure in which an ancient spirit and a modern technology are condensed, focused, in the figure of Hitler the Mediator, Hitler the Artist of the National Socialist State?[43] And indeed, this art (of the state, of politics) was, as Hans Jürgen

Syberberg has suggested, precisely *cinematic*: "Hitler created war only for the newsreels. To see his own epic Hollywood movie."[44] In a similar vein, Gilles Deleuze has noted that "up to the end Nazism thinks of itself in competition with Hollywood."[45] Anton Kaes confirms this point by drawing attention to Goebbels's "cinematic" vision—near the very end of the war—of the future of the Nazi state: "in a hundred years' time they will be showing a fine color film of the terrible days we are living through. Wouldn't you like to play a part in that film? Hold out now, so that a hundred years hence the audience will not hoot and whistle when you appear on the screen."[46]

Thus, it was perhaps only in terms of cinema, in terms of the "art" of "cinematic propaganda," that the imaginary "structure" of the Nazi cathedral of the future could be fully brought "to life," projected as a unifying aesthetic vision. If, then, the glass or crystal cathedral was figured as a kind of radio receiver, the cathedral of the future could only be a gigantic cinema hall, or rather, the Art of Cinema itself, projected onto the world as a mediated, aestheticized, and luminous totality, the ultimate *Gesamtkunstwerk*, a truly total cinema.[47]

The insistence on viewing this totalitarian art as merely a delusion, an ideological "screen" that hides the obscene truth of Nazi tyranny, misrecognizes the extent to which this tyranny is in fact based on just such an "aesthetic" of wholeness and mediation, in which the divisive repression of a technological, patriarchal law is to be overcome.[48] In other words, the violent, oppressive tyranny of Nazism is not based on an identification with the Technological Law of the Father, but on precisely a refusal of this identification: that is, on the denial of differences, of the technologies of difference. Difference, like technology itself, is either to be mediated into an aesthetic whole or excluded and destroyed. So too for the technology of cinema, which should, as Goebbels noted of *Triumph of the Will*, "overcome" the instrumentality of "mere propaganda" and "lift up the harsh rhythm of our great epoch to eminent heights of artistic achievement."[49] And even in this statement, one can discern the figures of a paternal technological (or instrumental) rationality "drawn upwards" by an eternal-feminine spirit in which it finds its mediation, its destiny, its home. It is only on the basis of this "aesthetic" mediation (of technology, of difference) that the tyranny and the horror of Nazism are made possible: that is, only in the name of an androgynous Mediator, through whom difference is denied.

3.
The Avant-Garde Technē and the Myth of Functional Form

By incorporating technology into art, the avantgarde liberated technology from its instrumental aspects and thus undermined both bourgeois notions of technology as progress and art as "natural," "autonomous," and "organic."

Andreas Huyssen, "The Hidden Dialectic"

Modernity has perhaps always defined itself within the dialectic between a utopian and a dystopian view of technology. In this dialectic, technology—whether it has been seen as a mere tool or as an autonomous process—has always been viewed in terms of an instrumental rationality. In aesthetic modernism, this utopian-dystopian technological dialectic remains applicable. Yet, at the same time, it is in modernist art that a different conception of technology begins to emerge, a conception in which technology is no longer defined solely in terms of its instrumentality, but also in aesthetic terms. Indeed, aesthetic modernism can itself be defined by this relationship: by both the aestheticization of technology and the technologization of art. From the late nineteenth century on, then, aesthetic modernism becomes the privileged site for the conjunction of technology and art. The two come together at the level of form, of technique.

If, however, the foregrounding of form or technique has been virtually axiomatic in discussions of modernist art, and of modern culture in general, the proper attitude toward this foregrounding has been the subject of considerable debate. This debate has, moreover, been divided largely along the lines of an art versus technology split, with the modernist emphasis on form or technique charged with either "aestheticism" or "technicism."

On the one hand, then, modernist art has been seen as a kind of formalist "aestheticism" that attempts to maintain, reconstitute, or even extend the autonomous, organic status of the aesthetic sphere against the instrumental rationality of the modern technological world. This characterization has, as Charles Russell notes, been particularly associated with "high" literary modernism:

> In effect, the modernist writers' aesthetic strategies announce the putative primacy of the creative consciousness and the literary work over the social domain. The work and its language thus not only create an apparently autonomous domain from which one critically views modern society, but also claim to offer new knowledge and a sense of authentic being not attainable in daily life.[1]

Yet the aestheticization of technique in "high" literary modernism is also, in a very strong sense, an aestheticization of technology; for, as Hugh Kenner has noted, it is precisely at the level of technique that "high" modernism shows the influence of modern technology: in the quick cuts between episodes that imitate the rapid movements of urban mass transport and in the complex narrative structures that resemble "one of those intricate feats of lever-and-wheel simulation."[2] In these aestheticized techniques, technology is, at least metaphorically, remotivated as a noninstrumental, aesthetic phenomenon, an artistic technique.

On the other hand, critics such as Renato Poggioli have observed the "technicism" of modernist or avant-garde art, the way that its emphasis on technique tends to technologize or instrumentalize aesthetic or spiritual values:

> [W]hat often triumphs in avant-garde art is not so much technique as "technicism," the latter defined as the reduction of even the nontechnical to the category of technique. "Technicism" means that the technical genius invades spiritual realms where technique has no raison d'être. As such it belongs not only to avant-garde art, but to all modern culture or pseudoculture. It is not against the technical or the machine that the spirit justly revolts; it is against this reduction of nonmaterial values to the brute categories of the mechanical and technical.[3]

At the point of Poggioli's attack here is the "technical-structural functionalism" of movements such as Productivism, which he sees as "neatly caricatured" in the "distant posterity" of Yevgeny Zamyatin's *We*, where "the timetable or general directory of the railroad" is viewed "as the unequaled and supreme masterpiece bequeathed them by this century" (p. 139). For Poggioli, then, what is at issue is not the aestheticization of technique or form but the technologization of aesthetic, spiritual values—their reduction, by a technical or formal rationality, to an instrumentalized, functional form.

The aestheticist and technicist positions serve to designate the poles of a dialectic that defines aesthetic modernism. In this dialectic, the aestheticist position attempts to posit an autonomous, utopian aesthetic space separate from the dystopian instrumentality of modern technicism. To this end, it aestheticizes technology. On the other hand, the technicist position attacks the autonomy of the aesthetic sphere in the name of a functional, technological utopia; it therefore attempts to technologize aesthetics.

The distinction between aestheticist and technicist positions parallels another frequently suggested division of modernist art: that between the formalist aestheticism of "high" modernism and the more functional social activism of the avant-garde. As Russell notes, "The avant-garde wants to be more than a merely *modernist* art, one that reflects its contemporary society; rather it intends to be a *vanguard* art, in advance of, and the cause of, significant social change."[4] Similar to Russell's distinction between modernist and avant-garde art is Peter Bürger's influential definition of the "historical avant-garde," a term intended to designate those avant-garde movements such as Futurism, Dada, Surrealism, and the left avant-garde in Russia and Germany during the 1920s.[5] Bürger, in fact, defines these movements by their opposition to the autonomous status of "art as an institution," which reaches its "terminal point" in the formalism of a bourgeois "Aestheticism, where art becomes the content of art":

> The European avant-garde movements can be defined as an attack on the status of art in bourgeois society. What is negated is not an earlier form of art (a style) but art as an institution that is unassociated with the life praxis of men. When the avant-gardistes demand that art become practical once again, they do not mean that the contents of works of art should be socially significant. The demand is not raised at the level of the contents of individual works. Rather, it directs itself to the way art functions in society. (P. 49)

Yet, as Bürger also notes, in order to make art as an institution practical, functional within society, these avant-garde movements drew an "essential element" from aestheticism:

> The avant-gardistes proposed the sublation of art—sublation in the Hegelian sense of the term: art was not to be simply destroyed, but transferred to the praxis of life where it would be preserved, albeit in a changed form. The

avant-gardistes thus adopted an essential element of Aestheticism. Aestheticism had made the distance from the praxis of life the content of works. The praxis of life to which Aestheticism refers and which it negates is the means-ends rationality of the bourgeois everyday. Now, it is not the aim of the avant-gardistes to integrate art into *this* praxis. On the contrary, they assent to the aestheticists' rejection of the world and its means-ends rationality. What distinguishes them from the latter is the attempt to organize a new life praxis from a basis in art. (Ibid.)

Thus, it is precisely on the basis of an art that, as in aestheticism, is defined as noninstrumental that the avant-gardes attempt to construct a practical, functional art. Yet, at the same time, as Andreas Huyssen observes, "technology played a crucial, if not the crucial, role in the avant-garde's attempt to overcome the art/life dichotomy and make art productive in the transformation of everyday life."[6] Indeed, for Huyssen, the influence of technology on the avant-garde is pervasive:

[N]o other single factor has influenced the emergence of the new avant-garde art as much as technology, which not only fueled the artists' imagination (dynamism, machine cult, beauty of technics, constructivist and productivist attitudes), but penetrated to the core of the work itself. The invasion of the very fabric of the art object by technology and what one may loosely call the technological imagination can best be grasped in artistic practices such as collage, assemblage, montage and photomontage; it finds its ultimate fulfillment in photography and film, art forms which can not only be reproduced, but are in fact designed for mechanical reproducibility. (P. 9)

Thus, as Huyssen observes, the historical avant-garde, in order to make art functional, attempts to technologize it—but this technologization, as Bürger points out, must not be reducible to a means-end or instrumental rationality. This paradoxical conjunction of a functional art and a noninstrumental technology might, therefore, be considered as the obverse of the aestheticist formula of a "nonpurposive purposiveness"; for, whereas Kant's notion of beauty stresses the *indefinite form* of aesthetic purposiveness, the impossibility of representing a purpose or end separate from the object as a whole, the avant-garde emphasizes a *specifically purposive or functional form*—a finite and technological rather than an indefinite and aesthetic form—that has, nevertheless, been detached from an instrumental rationality. Thus, it is in the *form* of a functional but nonpurposive technē that the avant-garde attempts to reconcile art and technology, "liberating" technology, as Huyssen notes, "from its instrumental aspects," while at the same time destroying the aura of an organic, autonomous art and making it functional, technological (in a noninstrumental sense).[7] The numerous attempts by the avant-garde to integrate technology and art—which can be summarized in the Bauhaus slogan "Art and Technology—A New Unity"—attest to this double-edged desire.

To a large extent, then, this avant-garde technē defines the historical avant-garde and distinguishes it from the more aestheticist position of "high" modernism.[8] Yet, although the technologized form of the artwork in the avant-garde technē has generally been regarded as a "functional form," it has also, perhaps more appropriately, been seen as a manifestation of a "machine aesthetic"; for the functionalist, technicist orientation of the historical avant-garde has been continually beset by an aestheticist impulse or, to be more precise, by a tendency toward "formalism." In this technological formalism, the form of the machine, and of a technological rationality, becomes detached from and takes precedence over its basis in instrumentality. Unlike in "high" modernism, however, this technological form or technique is not, in general, subordinated to an aesthetic ideal; it retains, for the most part, an identifiable status as technological. Thus, as has been noted, the avant-garde technē is constituted on the basis of form—a technologized form, a machine aesthetic—in which art and technology are each sublated, one in the other.

The construction of the machine aesthetic, then, begins on the two foundations of the technological and the aesthetic. It therefore seems reasonable, in analyzing this construction, to examine the two aesthetic forms that have most often been identified with the technological, and with modernism in general: architecture and film. The areas of architecture and design seem to offer a privileged space for considering the questions of mechanization and functional form, while film and photography seem most appropriate to the examination of technological reproducibility and the techniques of collage and montage. As will become apparent, however, this distinction is an artificial one.

The derivations of the architectural avant-garde that would come to be known under the headings of the "modern movement" or the "International Style" have been various: the English Arts and Crafts movement, French Rationalism, American and British engineering have all been seen as its predecessors. Yet the heritage of this avant-garde can also be traced, as can the term *avant-garde* itself, to the technicist utopianism of Saint-Simon, Fourier, and their followers, where the vanguard arts were to fulfill the didactic, utilitarian function of advancing the program of a rationalized, "engineered" society; thus, the institution of art was itself to become a tool, a technology, a practical means of engineering a rational, "scientific" utopia. Yet this technologization of art as an institution, which would be echoed by the avant-gardes of the twentieth century, was to manifest itself mainly at the level of the artwork's goals and content. The twentieth-century avant-gardes, on the other hand, would take a much greater interest in technologizing *artistic form,* in the aesthetic of the machine.

In the gap between these two avant-gardes, then, the development of a machine aesthetic comes about as artistic form is increasingly subjected to a technological rationality. In this process of rationalization, the form of art, like the institution of art, is supposed to become instrumentalized, functional. Artistic form, in other words, is also to become a technology, a means to an end. Not surprisingly, then, the technologization of architectural form was to borrow heavily from such "engineered" forms as bridges, factories, railway stations, exhibition halls, and, as Benjamin notes, the arcades.

Indeed, the nineteenth century was to see the status of the engineer rise to such a degree that, by 1889, the French architect Anatole de Baudot could observe: "A long time ago the influence of the architect declined and the engineer, *l'homme moderne par excellence,* is beginning to replace him."[9] Benjamin, in fact, finds that the dominance of the engineer begins with the building of the arcades, at which time "the idea of the engineer . . . begins to assert itself, and battle is joined between constructor and decorator, Ecole Polytechnique and Ecole des Beaux-Arts."[10] For Benjamin, the "constructional engineering" of the arcades, their use of "the functional nature of iron," would serve to liberate "constructive forms from art" (p. 161). These "constructive forms," then, are the very opposite of "artistic form," of an exterior style or ornamentation that hides its structural principles; they are, instead, purely "functional forms" that, like machines, are designed according to the rational, mathematical principles of efficient functioning. Their construction therefore becomes their "style," their form.

This combination of form and function is readily apparent in a structure such as the Crystal Palace, built for the Great Exhibition of 1851. There, the use of standardized, interchangeable parts allowed mass production to be brought to bear. Indeed, as Kenneth Frampton notes,

> The Crystal Palace was not so much a particular form as it was a building process made manifest as a total system, from its initial conception, fabri-

> cation and trans-shipment, to its final erection and dismantling. Like the railway buildings, to which it was related, it was a highly flexible kit of parts. Its overall form was structured around a basic 2.44-metre (8-foot) cladding module, assembled into a hierarchy of structural spans. . . . Its realization, which took barely four months, was a simple matter of mass production and systematic assembly.[11]

Although the scope of the Crystal Palace was exceptional—its surface area amounted to nearly a million square feet of glass—its use of prefabricated, mass-produced parts was not, for around that time, "cast-iron columns and wrought-iron rails, used in conjunction with modular glazing, had become the standard technique for the rapid prefabrication and erection of urban distribution centres—market halls, exchanges and arcades" (p. 33). Thus, the "functional nature" of materials such as iron, and later, reinforced concrete and steel, was not only a matter of their greater malleability and tensile strength, which allowed structures of increasing span and height; these materials were also functional in their capacity for standardization, for inexpensive, efficient mass production. The standardization of these components in a prefabricated structure like the Crystal Palace meant that its construction would also follow the model of mass production. The building process itself became organized into discrete, interchangeable component operations, a kind of assembly line for building production.

Indeed, the development of the assembly line in the nineteenth century is a direct result of the attempt to increase production through the standardization of the production process. This standardization, which is the basis of mass production, can be seen, in part, as a continuation of a tendency toward rationalization that begins in the Renaissance. Rationalization functions on the basis of a critical or analytic breakdown that abstracts and fragments its objects, reducing them to their component "elements," to their simplest possible forms, usually represented in quantitative or geometric terms. These elements can therefore be efficiently reproduced, assembled, organized, or otherwise manipulated. Yet standardization, as the figure of the assembly line suggests, also implies mechanization. The machine permits a more exact abstraction and quantification of particular productive operations than is possible with hand tools, allowing an increase in speed and precision while reducing the need for skilled labor. This mechanized rationalization corresponds, then, to what David Landes has called "the fundamental principle of industrial technology—the substitution of inanimate accuracy and tirelessness for human touch and effort."[12] For Landes, this principle combines two processes:

> (1) the fragmentation of the job into simple operations susceptible of being performed by single-purpose machines run by unskilled or semi-skilled hands; and (2) the development of methods of manufacture so precise that assembly became routine, in other words, the production of interchangeable parts. Only in this way could one change from a nodal to a linear flow; only in this way could one move the work to the workers at a predetermined pace, to be processed and put together by a series of simple, repetitive acts. The assembly line was thus far more than just a new technique, a means of obtaining greater output at less cost. In those branches where it took hold, it marked the passage from shop, however big and heavily equipped, to factory. (P. 307)

This mechanization and standardization of industry, which took place largely in the second half of the nineteenth century, would be seen as the work of the engineer, whose increasing prestige would be linked to the values associated with it: precision, efficiency, productivity, and the subordination of individuality in the interests of standardization; these became the traits not only of modern industry but of *l'homme moderne*. Indeed, an attempt would be made, through "scientific management," to extend these traits or values to workers in general. It is therefore hardly surprising that the leading proponents of scientific management, Frederick W. Taylor and his most prominent disciple, Frank Gilbreth, were both engineers. Two of Gilbreth's most important studies, in fact, concerned building processes, or "systems," as he called them: *Concrete System* (1908) and *Bricklaying System* (1909) recorded and analyzed the time and movements involved in the various aspects of ferro-concrete and brick construction. The principles of standardization and mechanization that had been applied to the iron and glass components of the Crystal Palace and to the production process in the assembly line were now extended to even human labor. With Henry Ford's introduction of the automatic assembly line and the Model T, the rule of a standardized, functional design and of an instrumental, "technicist" rationality appeared to be complete. Mechanization had, in Siegfried Giedion's phrase, "taken command."

This notion of a "triumphant" functionalist mechanization is the context into which Giedion and others, often including the avant-gardists themselves, have placed the historical avant-garde in architecture and the arts. Modern architecture, in particular, came to be seen within an

Ingenieurbauten tradition, which canonized the functionalist structures of the nineteenth-century engineer—bridges, exhibition halls, railway stations, the Crystal Palace, and the Eiffel Tower, as well as American grain silos and skyscrapers—as the forerunners of the modern movement. In this historical representation, the modern movement is portrayed as the progressive development of an architecture based on the rational, functional, and technological principles of engineering. In this development, the irrationality and superficiality of ornamental "styles" of design such as Art Nouveau are supplanted by the "styleless," mass-produced functional forms of an International Style. Indeed, as the term *International Style* suggests, the progressiveness of the modern movement is not presented simply as a matter of increasing rationality; it is also seen as a sociopolitical progressiveness that favors the "objective" interests of the masses over the subjectivity of bourgeois individualism. Thus, in contrast to the individualistic, handicraft-based ornamentation of bourgeois aestheticism, the standardized, functional objects of the International Style—by virtue of their susceptibility to mechanization and mass production—are to be universal and democratic, available to all. The history of the modern movement has been written largely in the terms of this trend toward a democratic, and often socialist, mechanization and mass production, where houses are to be "assembled as Ford assembles cars, on moving conveyor belts" and the "object" of design is to be "of no discernible 'style' but simply a product of an industrial order like a car, an aeroplane and such like."[13] Thus, Nikolaus Pevsner, one of the more prominent historians of the movement, can write that "The true pioneers of the Modern Movement were those who from the outset stood for machine art."[14]

This notion of an engineering or machine "aesthetic" would come to define avant-garde architecture and art. Thus, Vladimir Tatlin's monument to the Third International would be exhibited under a banner bearing the slogan "Engineers Create New Forms." In fact, the idea of the artist-engineer would be a favored theme of, among others, Adolf Loos, Le Corbusier, Berlin Dada, the Constructivists, and the Bauhaus (especially after 1923). Indeed, this trend toward engineering would be such that "art" would often be rejected altogether. As early as 1912, for example, Adolf Loos could observe: "Only a very small part of architecture belongs to art: the tomb and the monument. Everything else, everything that serves a purpose, should be excluded from the realms of art."[15] Benjamin put the matter similarly when he observed that with "the use of iron construction," architecture "begins to outgrow art."[16] It is in this sense that the aesthetic implications of "architecture" will be eschewed in favor of the more technicist terms of "building" and "construction."

A similar attitude is visible in the "fine arts," as is apparent in the slogan that served as the theme of the Berlin Dada Fair of 1920: "Art is dead! Long live the new machine art of TATLIN!" In accordance with this rejection of the "fine arts" in favor of the "engineered" forms of "the new machine art," many artists would turn to such "mechanical" arts as textile, industrial and graphic design, typography, and photography, photomontage, and film. The Soviet artists of the Inkhuk (Institute for Artistic Culture) would, for example, reject "pure painting" as a "speculative activity" and declare that "the last picture has been painted." Their attitude is summarized in a statement by Varvara Stepanova: "Technique and industry have confronted art with the problem of construction as an active process and not a contemplative reflection. The 'sanctity' of a work of art as a single entity is destroyed." Similarly, Liubov Popova would suggest that her paintings "must be considered simply as a series of preparatory experiments towards materialized constructions."[17] This movement toward what Osip Brik termed "Productivism" would in fact lead Stepanova and Popova to do design work for a textile factory, while Tatlin would go to work in a metallurgical factory.

The Productivist movement was, perhaps, the most extreme case of the trend toward technicism in the arts. There, the primacy of the engineered, functional form, of a technologization of the arts that would destroy the notion of autonomous art, received its most forceful advocacy. Yet, even in the instrumental "rationalization of artistic labor" advocated by the production-oriented First Working Group of Constructivists, an almost inevitable reification of technological form occurs, a tendency toward "pure" formalization that, although not strictly speaking aesthetic, will nevertheless lead back to a machine *aesthetic*, an aesthetics of functional form.

This return of a repressed formalism is, perhaps, endemic to the process of instrumental or purposive rationalization; for rationalization is always a matter of abstraction, and thus, of form. The abstractions of this formal rationality, on which technological form is based, are themselves a technology; so long as they remain in the service of an end external to themselves, they, like technology, remain functional, instrumental. Yet, as rationalization advances, the rational, technological form becomes not an instrument of the human mastery of nature, but an end in itself. Thus reified, the technological form is detached, or at least deferred, from its instrumental function.

This reification of technological form is the result of the extension of rationalization to the aesthetic sphere, to the sphere of aesthetic judgment. The nineteenth century, as Leo Marx has observed, had already begun to give voice to "a rhetoric of the technological sublime," as demonstrated in the following passage from 1832: "Objects of exalted power and grandeur elevate the mind that seriously dwells on them, and impart to it greater compass and strength. Alpine scenery and an embattled ocean deepen contemplation, and give their own sublimity to the conceptions of beholders. The same will be true of our system of Rail-roads."[18] This aestheticization of the machine's "power and grandeur" is frequently linked to an increase in humanity's power, to the ability to master "space and time." It is, therefore, easily extended to technology's utilitarian aspects, which even gain the aesthetic approval of God: "His works proclaim his preference for the useful to the merely imaginative, and in truth it is in such, that the truly beautiful or sublime is to be found. A steamer is a mightier epic than the Illiad,—and Whitney, Jacquard and Blanchard might laugh even Virgil, Milton and Tasso to scorn."[19]

At about the same time, Lothar Bucher could write that the impression produced by the Crystal Palace "was one of such romantic beauty that reproductions of it were soon hanging on the cottage walls of remote German villages."[20] In all of these examples, technology is aestheticized, but its beauty is not yet a matter of formal rationality or technological form. The engineer may create art, but art is not yet conceived in terms of engineering. And yet, there are, as Bucher notes, intimations of a new conception of art: the spectators of the Crystal Palace began "to realize that here the standards by which architecture had hitherto been judged no longer held good." This new architecture—with its emphasis on abstract, rationalized, and functional forms—is prophesied by another commentator of the times: "Mankind will produce a completely new architecture out of its period exactly at the moment when the new methods created by recently born industry are made use of. The application of cast iron allows and enforces the use of many new forms, as can be seen in railway stations, suspension bridges, and the arches of conservatories."[21]

In this passage, one can perhaps detect the connection of this new architectural aesthetic to the growing sense of modernity, which would affect not only architecture but all the arts—including that of the passage's author, Théophile Gautier. Yet, if this new aesthetic is based on the functional, technological forms of engineering, it must also be based on the abstraction and mathematization of a formal rationality. As soon, however, as the notion of functional form is seen in aesthetic terms, it becomes less a matter of function or technology than of a form that *looks* functional or technological. Thus, as the modern engineer/inventor is, throughout the latter half of the nineteenth century, elevated to the status of artist (e.g., Eiffel, Edison, Muybridge), the idea of a functional, technological form gives way to an *aesthetic* of functionality, of the machine—that is to say, to an aesthetic of formal rationality.

Perhaps the clearest example of this aestheticization of functional form—and of its dependence on an aesthetic of formal rationality (i.e., on an abstract, mathematical, and mechanical form)—occurs in the de Stijl movement, whose members, as Reyner Banham has noted, "seem to have the best right to be considered the true founders of that enlightened Machine Aesthetic that inspired the best work of the Twenties."[22] Central to the beginnings of this movement was a Neoplatonic, and indeed theosophical, opposition between nature and an abstract idealism; indeed, in the very first sentence of the first issue of the journal *De Stijl*, Piet Mondrian would proclaim: "The life of contemporary cultivated man is turning gradually away from nature; it becomes more and more an a-b-s-t-r-a-c-t life."[23] Even in his Paris Cubist years, Mondrian had been influenced by Theosophy's tendency toward abstract, universal forms. On his return to Holland in 1914, Mondrian was to fall even more heavily under the influence of the theosophical philosophy of mathematician M. H. Schoenmaeker, from whose *Beeldende Wiskunde* (Plastic mathematics 1916) he was to draw both the name and the tenets of "Neo-Plasticism" *[nieuwe beelding]*.[24] The abstraction of form and color emphasized by these tenets was to result in paintings, restricted to the primary colors, of abstract, rectilinear, and rectangular geometric forms. And although these "pure," mathematical abstractions are considered by de Stijl to be universal forms, opposed to the individual and the natural, they are also linked to modernity and to modern urban life: "The genuinely Modern artist sees the metropolis as Abstract living converted into form . . . that is why the metropolis is the place where the coming mathematical artistic temperament is being developed; the place whence the new style will emerge."[25] Implicit in this notion of the abstraction of the modern city is an appreciation of the abstract form of the machine, an appreciation that Theo van Doesburg would make explicit: "The machine is, *par excellence*, a phenomenon of spiritual discipline. Materialism as a way of life and art took handicraft as its direct psychological expression. The new spiritual artistic sensibility of the twentieth century has not only felt the beauty of the machine, but has also taken cognisance of its unlimited expressive possibilities for the arts."[26] Thus, the mathematically precise, impersonal abstraction of the machine form becomes a model for not only the "style" of the arts, but also for the movement away from individualized handicraft and toward mass industrial design. Indeed, as the architect J. J. P. Oud noted in his *De Stijl* article "Art and Machine," the "essence of the new style" lay in its combination of two trends: "a *technical-industrial* trend . . . which seeks to give aesthetic form to the products of technology" and a trend "by which art tries to achieve objectivity *[Sachlichkeit]* by means of reduction (abstraction)."[27] Both of these trends would find expression in Gerrit Rietveld's famous Red/Blue chair of 1917, illustrated in *De Stijl* in 1919.

Rietveld's chair can be taken to represent a shift away from the spiritual connotations of abstraction toward a more rationalist position that would eventually justify itself in terms not of the spirituality but of the functionality of abstraction. In it, as Banham notes, "The functions of the chair have been analysed, discriminated, reduced to their 'essentials'" (p. 190). These "essential" elements have then been assembled (and painted) in a way that highlights their discrete, abstract forms. In this elementarist mode, then, "functionality" becomes a matter of abstraction, of reducing functions to their logically, rather than spiritually, "essential," elementary forms. And it is precisely this abstraction that gives Rietveld's chair, for van Doesburg, the "silent eloquence of a machine." "Functionality," in other words, comes to be based on the abstract, rational form of the machine.

The trajectory of de Stijl from a spiritualized to a "functionalist" abstraction is remarkably similar to the roughly contemporaneous movement in Russia from Suprematism to Constructivism (as well as to the slightly later turning away of the Bauhaus from Expressionism toward *Neue Sachlichkeit*). Just as in Mondrian's abstract Neo-Plasticism, Suprematism and "nonobjective" art reject the representation of objects in favor of the "purely painterly plasticity" of abstract forms.[28] For Malevich, in particular, this nonobjective abstraction is explicitly spiritual. Yet the emphasis on abstract, geometric forms in his painting suggests a connection to mathematics, science, and technology that others would make considerably more of. And even Malevich, although he rejects the Futurist fascination with machines as "objectism," suggests that it is the machine that has opened the possibility of new, spiritualized abstract forms:

> The new life of iron and the machine, the roar of motorcars, the brilliance of electric lights, the growling of propellors, have awakened the soul, which was suffocating in the catacombs of old reason and has emerged at the

> intersection of the paths of heaven and earth. If all artists were to see the crossroads of these heavenly paths, if they were to comprehend these monstrous runways and intersections of our bodies with the clouds in heaven, then they would not paint chrysanthemums. (P. 126)

Even in these early statements of 1915, Malevich seems at times to suggest that it is largely their noninstrumental or aesthetic quality that distinguishes "pure," nonobjective forms from the forms of technology, of "utilitarian reason."[29] By 1919, Malevich would propose placing "all the arts—painting, color, music, constructions—under the umbrella of 'technological creation.'" The basis of this creation is to be what Malevich calls "the fifth dimension": "economy."[30] It must be noted, however, that for Malevich, the concept of economy seems to remain a notion of formal economy and rationality rather than of practical economics. His Suprematist "architectons," for example, are more experiments in abstract form than practical models for building. In this respect, they are much akin to Tatlin's more explicitly machine-influenced model for a monument to the Third International. Indeed, similar to Malevich's emphasis on "economy," Tatlin will, in reference to his project for the Third International, observe the "severity" associated with "materials like iron and glass," which opens the possibility of "uniting purely artistic forms with utilitarian intentions."[31] Thus, the abstraction and rationalization of Suprematist and nonobjective forms come to be seen—precisely on the basis of their formal "economy" or "severity"—as models for the standardized, mechanical forms of utilitarian mass production. These implications are spelled out quite clearly by Malevich's student and collaborator, Ilia Chashnik, in this description of the Department of Architecture and Technology at Malevich's Vitebsk art school:

> The constructions of Suprematism are blueprints for the building and assembling of forms of utilitarian organisms. Consequently, any Suprematist project is Suprematism extended into functionality. The Department of Architecture and Technology is the builder of new forms of utilitarian Suprematism; as it develops, it is changing into a huge workshop-laboratory, not with the pathetic little workbenches and paints in departments of painting, but with electric machines for casting, with all kinds of apparatuses, with the technological wealth of magnetic forces. [This department works] in concert with astronomers, engineers, and mechanics to attain a single Suprematism, to build organisms of Suprematism—a new form of economics in the utilitarian system of contemporaneity.[32]

Here, one can observe a curious "economic" conflation, a kind of "dual economy" that would be similarly apparent in El Lissitzky's juxtaposition of Suprematist elements and a snowplow on the cover of the journal *Veshch.* This "dual economy" is also the equivalent of the two trends that Oud saw in de Stijl: in one, economy is a matter of formal reduction or abstraction—an economy or simplicity of form; in the other, economy is a functional, technological, or industrial matter, to which a formal economy (an aesthetic ideal?) is to be applied. It should be clear that formal economy is not in this case determined by functional, economic considerations. Rather, abstract, economical form is seen as the basis for economical, efficient functioning—in both architecture and technology.

Yet if, in the case of Chashnik (and Lissitzky), formal economy remains an abstract ideal that is to be "applied" to the practical, industrial arts, it is worth noting that even the Productivists, while disavowing formalist aestheticism and idealism, do not completely abandon the formal rationality and abstraction of their earlier "laboratory" work. Rather, formal rationality comes to be equated with "scientific" analysis. It is this "science" that, along with the "science" of dialectical materialism, will inform the utilitarian "technology" of industrial production or construction. Indeed, the abstract, geometric forms of this "scientific" rationality become the technology, the technique, through which function is to be implemented.

This association of rationalized, geometric forms with technology and machines often leads to the assumption that such forms are necessarily—that is to say, essentially—functional. As noted earlier, the prestige accorded to engineering and technology encourages this reification of a machine *aesthetic,* of technological form as an end in itself. Similarly, the supposedly functional process of scientific analysis is itself reified—abstracted and rationalized into a "purely" formal rationality. Scientific analysis is, after all, a matter of reducing objects under analysis to their simplest, most basic components, abstracting their fundamental forms, elements, or characteristics. Because it is "scientific," then, this process of formal reduction and abstraction is also presumed to be functional. It is the name of "science" that provides the guarantee that reducing objects to basic, elementary forms or elements is an inherently functional, "scientific" process.

As already noted, it is the Productivists who take these principles of "scientific method" and "functional form" to their logical extreme. As Brik notes, "The basic idea of productional art [is] that the outer appearance of an object is determined by the object's economic purpose and not by abstract, aesthetic considerations."[33] Yet, in order to achieve the "object's economic purpose" in an "economical" manner, form must itself become "economical." Thus, for the Productivists, just as for the Constructivists and de Stijl, "functional form" becomes a matter of *formal* economy or simplicity, of nonornamental forms. As Banham puts it, the idea "that to build without decoration is to build like an engineer, and thus in a manner proper to a Machine Age," becomes "vital" to avant-garde notions of design (p. 97).

This notion that economical, nonornamental forms are inherently "functional" is based, as already noted, on an analogy in which the formal rationality of science becomes the model for technological, functional forms. Even as the Productivists oppose the notion that form should be determined by "aesthetic" criteria, they substitute the idea that form should be determined by a "scientific aesthetic"; the principles of formal rationality and economy that are associated with the "scientific method" become, in other words, "aesthetic principles" to be applied to objects or products. The Productivist Nikolai Punin, for example, sees science as the underlying principle of Soviet culture:

> If modern man wants to assimilate fully all the forces affecting the creation of this or that work of art, he must approach the work by studying and analyzing it by means of scientific method. Science is not a symptom but precisely a principle. . . . We do not strive for science to develop and prosper in our world; we strive primarily in order that our whole world view, our social structure, and our whole artistic, technological, and communal culture should be formed and developed according to a scientific principle.[34]

For Punin, this analytic, scientific principle is precisely a matter of efficiency or economy, and just as these concepts are applied in mechanization and mass production, they must also govern artistic production:

> In the new arrangement of European society—which has not yet come about, but which is in evidence—man must as far as possible economize his energy and must in any event coordinate all his forces with the level of modern technology. . . . Insofar as the artist strives to approach the machine in his creative process, insofar as he wishes to regulate, to mechanize his forces in accordance with the contemporary order, with the contemporary trends of progress, mechanization becomes a stimulus for creating a new artistic culture. (P. 175)

Indeed, Punin will actually go so far as to reduce the creative process to a mathematical formula.[35] Similarly, Popova calls for an "infallible" formula, "like the formula of a chemical compound, like the calculated tension of the walls of a steam boiler, like the self-confidence of an American advertisement, like 2 × 2 = 4," that would provide "a single, integrated approach to evaluating the facts of everyday life."[36] For Popova, work in "material design" is possible only through these "precise calculations" or formulas, which reduce the world to its rationalized, "essential" elements or forms, which can then be efficiently, economically assembled. The model, of course, is precisely that of the engineer or of the assembly line, where the "scientific" principles of formal rationality are applied. As a "principle," however, this formal rationality or economy is no longer subordinated to function; the rationalized form, the lack of "artistic" ornament, becomes an end in itself. The concept of functional form, then, comes to be seen not so much as a matter of functionality, of an instrumental rationality, but of a formal rationality that comes to stand in for, to represent or simulate, functionality. Indeed, it is precisely this simulacral status that allows the avant-garde technē to be constituted: the rationalized forms of modernism come to be seen as "functional forms" not because they are functional or technological but because they represent functionality and technology. A geometric textile design is not, after all, more functional than a flowery print; it simply *looks* more functional. Thus it is that the "reconciliation" of art and technology in the avant-garde technē is based on the "look" or "form" of technology, on a technological or machine *aesthetic.*

In the avant-garde technē, then, one can observe the beginning of a shift in the conception of technology. A notion of technology as instrumental, as the functional application of science, begins to give way to a conception that sees technology as a matter of form, of representation. And just as representation, for the avant-gardes, becomes increasingly allegorical, arbitrary, simulacral, so too does technology. Indeed, this new conception of technology can be designated as *simulacral* or *techno-allegorical.* In it, technology becomes a matter of forms that, analyzed and abstracted from what they represent, can be assembled, constructed, into any order, with virtually any meaning or purpose. Technology has perhaps always carried this unsettling "allegorical" sense of artifice and construction, of representation and simulation. This is, no doubt, the sense that allows Heidegger to see the link between modern technology and art. Yet, as Heidegger is acutely aware, what defines technology as modern is precisely the subordination of this techno-allegorical sense to a notion of technology (and science) that sees the world as "standing reserve" to an instrumental conception of technology. Indeed, in taking up the concept of "functional form," the historical avant-gardes reaffirm, sometimes quite explicitly, a Renaissance notion of modernity that sees science and technology as the means to achieve a rational knowledge of and control over nature. Thus, these avant-gardes frequently seem unaware, in their statements, of the shift in the conception of technology that has taken place at the level of their artistic practice.

In the avant-garde technē, then, this transition from an instrumental conception of technology to a techno-allegorical one is not yet fully recognized. Indeed, the avant-garde technē defines itself precisely in terms of the conjunction of these two notions of technology, a conjunction that can only be maintained through a representational "trick": the myth of "functional form." Yet, even as this myth enables the avant-garde technē's paradoxical juxtaposition of a noninstrumental technology and a functional aesthetic, it is necessarily based on the arbitrary relation between the sign and what it represents, on the detachment of formal rationality from instrumental rationality, on the deferment of technological form from technological function. And it is precisely this deferment that allows the rationalized forms of technology to become allegorical rather than instrumental—that allows technological form to become, in other words, a matter of technological reproducibility, and of the techniques of montage/assembly that are associated with it.

Yet, even as the conception of technology in the avant-garde technē comes to be regarded in terms of technological reproducibility and montage techniques, this conception remains subject to the myth of functional form. That is to say, technological reproducibility and montage techniques are themselves represented as "functional," a representation that is once again founded on the analogy to the engineer and the assembly line. Technological reproduction's reduction of the artwork, and the world, to the level of disenchanted images, signifiers shorn of aura, is seen as analogous to the engineer's analysis and abstraction of rationalized, standardized forms. These rationalized images/forms can then be assembled, constructed, by the allegorist/engineer, by the *monteur.*

And indeed, in the avant-garde technē, this conjunction of technological reproduction and mass production is figured precisely in the terms *montage* and *monteur.* Hannah Höch, for example, makes this connection clear when she observes that, in photomontage, "Our whole purpose was to integrate objects from the world of machines and industry in the world of art."[37] Moreover, as Raoul Hausmann notes in his explanation of the term *photomontage*, this connection of technological reproduction to engineering and industrial technology is not simply iconographic: "We decided to call this process *photomontage* because this term expressed our aversion to playing the part of the artist. Regarding ourselves as engineers (hence our preference for workmen's overalls), we meant to construct, to *assemble*, our works."[38] John Heartfield would, in fact, be known as the Monteur Heartfield not only for his photomontages, but for his "engineer-like" attitude and dress. A similar linkage of montage to engineering and industrial technology is also apparent in the Soviet Union, as Gustav Klutsis observes: "Photomontage . . . is closely linked to the development of industrial culture and of forms of mass cultural media."[39] Nor is this linkage restricted to photographic montage, as Sergei Eisenstein makes clear: "What we need is science, not art. The word *creation* is useless. It should be replaced by *labor*. One does not create a work, one constructs it with finished parts, like a machine. *Montage* is a beautiful word: it describes the process of constructing with prepared fragments."[40]

The centrality of montage and technological reproduction to film would, of course, lead many avant-gardists—from László Moholy-Nagy to Benjamin—to see film as the exemplary form of modern, technologized "artistic" production; for, even more than modern architecture, film is capable of exhibiting its technological, constructed status, of making this constructedness, its montage, the basis of its design, its form. Thus, film form does not so much follow function as it does the techniques of montage. Montage, in this case, is obviously not simply film editing in general, just as "constructivism" does not refer to construction generally. Rather, "montage," as used by the avant-gardes, is designed to emphasize, like the notion of film that is based on it, the fragmentary, allegorical, and, indeed, technological status of its own construction—in contrast to the symbolic unity and aura of the traditional work of art.

At the same time, however, many avant-garde film theorists see montage (and technological reproduction) as inherently functional. This subordination of a techno-allegorical sense of montage to a notion of montage as technological in a functional sense can only be accomplished through reference to the model of a scientific-technological rationality, by conflating montage and technological reproduction with the techniques of modern science and mass production. This recourse to science recalls the approaches of such cinema pioneers as Marey and Muybridge, whose work with photographic and cinematic montage was in fact used in scientific studies analyzing motion, rather than in bourgeois artistic representation. Certainly, the avant-garde film theorists saw themselves in this tradition, as employing filmic montage scientifically: to analyze, to take apart, the world. They, however, extend this notion of cinema's scientific, functional basis to include its effect on an audience. To this end, they attempt, using the model of "scientific-technological" analysis and rationalization, to reduce film to its basic, calculable laws, elements, forms, or, as Eisenstein puts it, "units":

> Don't forget that it was a young engineer who was bent on finding a scientific approach to the secrets and mysteries of art. The disciplines that he had studied had taught him one thing: in every scientific investigation there must be a unit a measurement. So he set out in search of the unit of impression produced by art! Science knows "ions," "electrons," "neutrons." Let there be "attractions" in art. Everyday language borrowed from industry a word denoting the assembling of machinery, pipes, machine tools. This striking word is "montage" which means assembling, and though it was not yet in vogue it had every qualification to become fashionable. . . . Thus was the term "montage of attractions" coined.[41]

Eisenstein's tendency toward "scientism"—what Jacques Aumont has called his "will to science"—is well known and is obvious here.[42] Indeed, Eisenstein would soon abandon the "montage of attractions" for the even more scientistic notion of a montage of shocks or stimuli, conceived on the model of reflexology. As Aumont notes, however, what is maintained throughout the numerous shifts and turns of Eisenstein's terminology over the years is his belief in the possibility of a science that would define the formal laws of cinema. For Eisenstein, this "science" is always a matter of functionality, of efficacy, that is, of film's ability to serve as an efficient, instrumental technology, as "a tool to exert an influence on people."[43] In other words, Eisenstein sees his attempts to discover the "scientific" basis of the cinema, to determine its "units of measurement," as enabling a "calculation" of the spectator. Thus, Eisenstein's "will to science," his continual effort to find a scientific, calculable grounding for his formal theories, serves to provide a "rationale" for the connection of form and function, a "legitimation" for the functionality of montage.

Eisenstein was not alone in associating montage with filmic efficacy or functionality. Lev Kuleshov, for example, would note that "the essence of the cinema, its method of achieving maximal impression, is montage."[44] Similarly, Dziga Vertov gives the following "word or two about efficiency and montage":

> The efficient should coincide with the beautiful. Take the racing automobile, for example, meant to go so many miles per hour maximum and made in the shape of a cigar. Here is both beauty and efficiency. We feel that one must not think only of the beauty of this or that shot. First of all, one must proceed from considerations of efficiency. This means that one adapts the angle from which a given object can best be seen.[45]

For Vertov, in fact, this technological efficiency of montage defines what is cinematic, what he calls "kino-eye." Thus, "a 100 percent film-object" can only be produced by kino-eye's "scientifically experimental method" of montage, a method that yields a "visual equation, a visual formula expressing the basic theme of the film-object in the best way."[46] For these theorists, as for the avant-gardes generally, montage appears as an inherently functional form. And, as in modernist architecture and design, this functionality is based on a formal rationalization, on a "scientific" analysis designed to reduce film to its simplest, most economical, and most basic elements, forms, and units, which can then be efficiently, mechanically assembled—as in a factory. Thus, the functionality of montage is not *simply* a matter of efficacious assemblage—for montage itself is seen here not only as a matter of assembly, but also as a matter of reduction, of "cutting."

"Cutting" should serve to refer here both to the temporal cutting that determines shot duration and to the "cutting out" of space through framing (i.e., through technological reproduction). Indeed, Eisenstein makes it very clear that he considers framing to be an important aspect of montage; and for him, framing is precisely a matter of "cutting out," of "hewing out a piece of actuality with the ax of the lens."[47] Cutting, then, involves a selection and highlighting of a particular element, form, or fragment. It also involves an abstraction or reduction of film form. The piece or fragment *(kusok)* thus abstracted—or "prepared," as Eisenstein says—comes to serve as the "compositional unit" of montage, the equivalent of the standardized part on the assembly line. Similarly, the process of cutting is seen as analogous to the analytic, "scientific" method by which the engineer determines the appropriate, standardized "functional form." An effective, efficient assemblage is supposed, as in a factory or machine, to be based on this "scientific-technological" rationalization of form, in which the formal economy of the standardized part coincides with its functional efficacy. Thus, the idea of a functional montage is only possible to the extent that montage is seen as reducible, at least in theory, to a "scientific," rationalized, calculable basis or formula—whether of units, as for Eisenstein, or of facts or intervals, as for Vertov. In other words, by analogy to the factory or machine, the cutting of montage is thought to enhance filmic efficacy by reducing film to its minimal, rationalized forms, its units or formulas, which are seen as inherently functional.

This scientific-technological analogy serves, then, to link the "cutout," technologically reproduced forms of montage and the functional forms of engineering and mass production; through it, both film and photography, on the one hand, and modernist architecture and design, on the other, become subject to the concept of functional form. This analogy also serves to explain the avant-garde's interest in pseudoscientific theories of acting, in which actors and performance are reduced to a series of codified types and "biomechanical" gestures. It explains, as well, the avant-garde fascination with attempts by "scientific management"—often based on photographic and filmed studies—to reduce labor to its most efficient, rationalized forms. Similarly, it provides a theory for the striking analogy of forms, noted by Sigfried Giedion and others, between modernist painting and the photographic motion studies of Muybridge and Marey.[48] And last but not least, it provides a context in which the left avant-gardes' often-noted but somewhat surprising admiration for not only American factories and skyscrapers, but films and advertising, can be understood: not only "functional" designs but technologically reproduced cultural forms are evaluated in terms of the functionality of "scientific" mass production.

Thus, the analogy that equates montage and technological reproduction with mass production can be seen as the "crux" of the avant-garde technē: it "crosses" representational efficacy with the functionality of mass production. In, for example, Eisenstein's attempts to "calculate" the spectator, the efficacy of a film is equated with the functional form of the mass-produced product. Thus, for Eisenstein, the "functionality" of the film product is based on scientific-technological principles, on its ability to rationalize formally economical, standardized basic units, which can then be efficiently assembled and conveyed through montage. This notion of representational efficacy or functionality presumes, however, that rationalization at the level of production will also be effective at the level of reception, of use. Thus, for filmic representation to be "functional" at both of these levels, representation must itself be "seen" as scientific, technological.

Vertov's notion of "kino-eye," for example, proclaims that a truly functional representation is only possible through the scientific-technological means of the camera and montage. Vertov, in fact, constantly contrasts this "scientific-technological vision" to that of the human eye: "Our eye sees very poorly and very little—and so men conceived of the microscope in order to see invisible phenomena; and they discovered the telescope in order to see and explore distant, unknown worlds. The movie camera was invented in order to penetrate deeper into the visible world, to explore and record visual phenomena."[49] Kino-eye, then, "gathers and records impressions in a manner wholly different from that of the human eye."[50] It is, moreover, an explicitly technological mode of representation: "I am kino-eye, I am a mechanical eye. I, a machine, show you the world as only I can see it" (p. 17). Kino-eye is, then, *more* than just filmic representation. In contrasting it to those "artistic" representations that use film merely to "copy" the vision of the human eye, Vertov foregrounds its technological basis—its technique and its montage:

> Kino-eye makes use of every possible kind of shooting technique: acceleration, microscopy, reverse action, animation, camera movement, the use of the most unexpected foreshortening—all these we consider to be not trick effects but normal methods to be fully used.
>
> Kino-eye uses every possible means in montage, comparing and linking all points of the universe in any temporal order.[51]

This desire to "expose technique," to make this "technological vision" the basis for film form, has clear roots in Russian Formalism, and Vertov's insistence here that these techniques are not merely "trick effects" is undoubtedly, in part, a response to the charges of formalism that were frequently leveled against him. Yet, throughout Vertov's writings, he views kino-eye as a method, one practiced not for its own, formalistic sake, but in order to discover and explore—to represent—a real truth:

> Not kino-eye for its own sake, but the truth through kino-eye, that is, kinopravda. Not filming "unawares" for its own sake, but to show people without masks, without makeup; to catch them with the camera's eye in a moment of nonacting. To read their thoughts, laid bare by kino-eye.
>
> Not "trick" effects, but the *truth*—that's what matters in kino-eye work.[52]

Thus, kino-eye's technological vision, its exposure of technique, finds its grounding in Vertov's unwavering emphasis on "facts," on "life," on the "nonacted," on the "documentary": "kino-eye = the kino-recording of facts." Recorded facts become, then, the basic units of kino-eye representation. Their status as "true" representational units is, however, dependent on a technological vision or representation, on kino-eye's capacity to represent life more perfectly, more functionally, than the human eye.

The relation between these "kino-facts" and the technological vision of kino-eye that brings them into existence is tautological, much like the tautology that allowed Renaissance science both to be founded on and make possible the "correct" representation of the world by mathematical forms or units. And indeed, Vertov, much like Eisenstein, often figures kino-eye's representation of "facts" as a "scientific" method: "The kino-eye method is the scientifically experimental method of exploring the visible world—a) based on the systematic recording on film of facts from life; b) based on the systematic organization of the documentary material recorded on film."[53]

Yet the relation between the forms of a photographic or filmic representation and what they represent is not, as in "scientific" representation, mathematical; it is visual. These forms, in other words, *look* like what they represent, they *simulate* it. Unlike the forms of mass production, then, which are defined by their ability to be scientifically reduced to simplified geometric forms or mathematical units, the technological reproduction of these images as "units" of representation is based on visual, contextual criteria. Indeed, these images can only be seen as "units" of representation metaphorically, by analogy to the functional forms of mass production. As such, they have some functionality at the level of production, but not at the level of their reception or use. It is, in fact, precisely this distinction between production and use that is disregarded in the avant-garde analogy between "scientific" mass production and technological reproduction (and montage).[54]

Put simply, this analogy confuses a productional functionality with a receptive, representational efficacy. It assumes that if technologically reproduced representations such as film can, like products of an assembly line, be efficiently assembled from scientifically abstracted, formally economical units, they are necessarily functional. And indeed, this process of rationalization may, as in the assembly lines of the Hollywood studios, allow the film product to be rationalized into more or less standardized "units" and assembled according to certain formulas. Through this process, it is possible that the film product may be more efficiently, more economically produced. This process of rationalization does not, however, necessarily make the film more functional in its effect on its audience. It could only do so if filmic representations were indeed reducible, at the level of both production and reception, to mathematical forms or units and formulas for combining them.

But then, this confusion between the functionality of production and the functionality of use occurs even in regard to mass production itself.

In the Soviet Union, at least, this confusion can perhaps be seen as a consequence of a general unfamiliarity with industrial technology. It is certainly a consequence—and this point can be taken to apply to the avant-gardes in general—of a lack of understanding of mass production within capitalism, and particularly in the United States. There are, however, several levels to this confusion. The first is a basic confusion between the functionality of the production process and the functionality of the product, a confusion that begins when the assembly of machines is taken as a model. In the case of the machine, there is perhaps some justification for the idea that the rationalization of the production process—the analysis and reduction of extraneous form, the use of interchangeable parts—leads to a more efficient, functional product: machines with more simple parts may in fact be more functional. For the most part, however, the analyzing of a product into basic, standardized parts—while it may enable more efficient production—does not necessarily make the finished product more functional in use (although it should, presumably, make it cheaper). A second confusion here, based partially on the first, is the idea, taken up in both production design and modern architecture, that a more functional product—and here again functionality of use is conflated with functionality of production—will necessarily result from a formal economy, a lack of ornament, in design. Yet this notion of economy in design—typified in the modernist reliance on abstract, geometric forms—does not necessarily lead to products that are cheaper to produce, and it especially does not make them more functional to use, as a glance at Rietveld's chair, a Bauhaus teapot, or a Productivist textile design will make clear.

The case is even clearer in modernist architecture, which—despite its slogans proclaiming that houses should be "machines for living in" to be "assembled as Ford assembles cars"—is often no more functional in its construction or its use than any other particular style.[55] Indeed, Buckminster Fuller, noting that the architects of the modern movement had little "knowledge of the scientific fundamentals of structural mechanics and chemistry," observes: "The International Style 'simplification' then was but superficial. It peeled off yesterday's exterior embellishment and put on instead formalised novelties of quasi-simplicity, permitted by the same hidden structural elements of modern alloys that had permitted the discarded *Beaux-Arts* garmentation."[56] Along similar lines, Reyner Banham has noted that the formal economy of modernist architecture was based more on an aesthetics of basic, geometric forms than on a serious functionalism. In fact, he observes, "if the architecture of the Twenties is regarded in the purely materialistic terms in which it is commonly discussed, much of its point will be lost. Nowhere among the major figures of the Twenties will a pure Functionalist be found, an architect who designs entirely without aesthetic intentions."[57] Rather, the "functional form" of modern architecture was often merely a simulation of functionality; the machine aesthetic was based on a representation of technology and functionality that was largely "symbolic":

> Emotion had played a much larger part than logic in the creation of the style; inexpensive buildings had been clothed in it, but it was no more an inherently economical style than any other. The true aim of the style had clearly been, to quote Gropius's words about the Bauhaus and its relation to the world of the Machine Age: . . . to invent and create forms symbolising that world. (P. 321)

These forms only *represent* or *simulate* the technological functionality of modern mass production. They do not so much follow function as borrow the idea of it. Indeed, whatever efficacy these forms may have is not based on a functional but on a simulacral or allegorical notion of technology and form. They are, in other words, effective as representations precisely because their "technological" form has been "cut," separated, from technological function, thus making them available for "assembly" at a symbolic level, as representational fragments. Yet, unlike the rationalized forms of mass production, these fragments are not simply standardized forms, calculable units derived from a rational, scientific analysis in which form is always subordinated to the principles of economy, efficiency, and functionality. They are not so much functional as simulacral forms, derived not so much from the rational principles of modern science and mass production as from the allegorical principles of technological reproduction—understood here both in the usual, Benjaminian sense, as reproduction by a technological means (such as film), *and* as a reproduction of technological form (as in modern architecture). But then, Benjamin had himself suggested the similarity of these two senses: "In the nineteenth century [the development of the forces of production] emancipated constructive forms from art. . . . Architecture makes a start as constructional engineering. The reproduction of nature in photography follows."[58] Thus, technological reproduction, in either sense, liberates or cuts out "constructive," technological forms. An important question here is whether Benjamin sees this cutting and these forms as technological in a functional or an allegorical sense. And although Benjamin frequently draws the analogy between mass production and technological reproduction, there is reason to think that he figures technological reproduction not so much in terms of functionality, but instead sees mass production in terms of allegory.

But whatever Benjamin may or may not have thought, the very fact that he can be read, or used, in more than one way suggests that the representational efficacy of his text cannot be reduced to unitary formal terms or formulas. And certainly, Benjamin's own use of montage technique, although it might be seen as analytic, can hardly be described as mathematical or scientific. A similar observation might be made in regard to many avant-garde works. For example, the films and even the writings of Vertov and Eisenstein display a fascination with formal experiment, with the various techniques of cutting and montage at their command, that far outstrips the functional, scientific-technological pretensions of their explicit pronouncements. Thus, Aumont's comment on one of Eisenstein's essays might as easily apply to *Strike* or *Potemkin* or to Vertov's *Man with a Movie Camera* or *Enthusiasm*: it conveys, Aumont says, "the impression of a machine whose motor is racing, of a calculation that is overwrought, of a gratuitous performance."[59] This, clearly, is not a technology or a form that can be called "functional," but one imbued with the irreducible surplus of simulation, of form, of technique. A similar, if not quite so exuberant, surplus can, as Banham has noted, be found in the technological form of the modernist building: "It is clear that even if it were profitable to apply strict standards of Rationalist efficiency or Functionalist formal determinism to such a structure, most of what makes it architecturally effective would go unnoted in such an analysis" (p. 323). Here again, efficacy of reception or use is not simply reducible to a functional form; the technological form of modernist architecture carries a formal, representational excess that Banham refers to as "aesthetic," but that might, more precisely, be called "formal" or "stylistic." And it is only by ignoring this excess that the avant-gardes can reduce the techno-allegorical form of filmic montage or modernist architecture to the status of the functional technological forms of a "scientific" mass production.

The avant-garde denial of this formal, simulacral surplus is based on its misunderstanding of the role of representational forms, of a simulacral technology, within capitalist society. The avant-garde model for a functional mass production was derived from the "scientific," production-based notions of Taylor and Ford. Yet, as American manufacturers, including Ford, would soon discover, production based on standardized, functional forms does not necessarily ensure efficacy at the level of reception, of consumption. Indeed, in their notion of a mass production based on the concept of a standardized functional form, the avant-gardes fall victim to what might appropriately be called the "Fordist fallacy," in reference to Henry Ford's theories of standardization. It was these theories that led him to continue to produce the Model T, even after Chevrolet's huge successes in introducing styling and color to mass-produced automobiles. The Fordist fallacy, then, subordinates consumption, like mass production, to the principles of economy, efficiency, and functional form. The efficacy of reception or consumption is, in other words, assumed to be a matter of the economy and utility of the product, of the product's efficiency in fulfilling rational needs. Yet, in both this Fordist fallacy and the concept of functional form on which it rests, it is not so much functionality that is the end or goal; rather, it is a formal, scientific-technological rationality that becomes its own end, its own mechanical aesthetic.

Subjecting consumption or reception to this formal rationality means that representation must also be rationalized. Technological reproduction and montage are, in fact, the result of this attempt to rationalize representation. Rationalized forms, however, always contain an excess that cannot simply be reduced to functional forms and rational needs. This excess is, in other words, not a matter of function but of a visual representation or copy; that is, the relation between the form of a product and its function is more or less arbitrary, allegorical; it is based on a technological simulation of functionality. And it is this simulacral technology that will determine the form of the mass-produced object, or rather, its *style.*

Indeed, with Ford's introduction of the Model A in 1927, and the rise of independent industrial designers such as Loewy, Bel Geddes, and Teague shortly thereafter, the dictum of "form follows function" gives way to a new slogan: "styling follows sales." In this slogan, capitalist marketing and product design acknowledge what the avant-gardes do not: that the nonutilitarian, simulacral aspect of consumption, its style, cannot be simply subordinated to the functional forms of production. It is perhaps worth recalling here Jean Baudrillard's critique of use-value and utility in *For a Critique of the Political Economy of the Sign*; utility, for Baudrillard, is merely an alibi for a system of exchange and consumption "*where the commodity is immediately produced as a sign, as sign value, and where signs (culture) are produced as commodities.*"[60] And certainly, the marketing emphasis on product styling (as opposed to form) is based on the recognition that utility is accessory to a consumption whose basis is the homology between the commodity form and the form of the sign; for it is the style of the product that, through its reproduction or simulation of certain values (e.g., functionality, technology), makes them liable to exchange and consumption.

In capitalist consumption, then, style does not simply represent certain preexisting values. Values, in any case, are always already represented; they can only be known through specific forms. Style, rather, simulates or technologically reproduces these value-forms, turning them into "objects" of exchange whose significance depends on their context. Thus, the automobile fin, in simulating the speed and modern technology of the rocket or jet fin, retains, for the most part, only a certain connotation of up-to-dateness, of stylishness. And indeed, in capitalist consumption, the value of style begins to become a purely differential matter, a matter of stylishness, for which the model, as Baudrillard points out, is fashion: "neither the long skirt nor the mini-skirt has an absolute value in itself—only their differential relation acts as a criterion of meaning. The mini-skirt has nothing whatsoever to do with sexual liberation; it has no (fashion) value except in opposition to the long skirt" (p. 79).

Baudrillard is quite clear, moreover, that the process of simulation that produces these "objects" of fashion, of exchange, is linked to a shift in the conception of technology, to a "mutation of [a] properly industrial society into what could be called our techno-culture" (p. 185). In this "techno-culture," the "rationalization" of consumption has turned on itself, has begun to consume its own tail. Any end or value above or beyond this cycle has been discarded, liquidated. Style has become its own end, its own value. Here too, the technology of mass production can no longer be seen simply as functional, but in terms of a truly simulacral technology—allegorical and arbitrary—where technological reproduction and montage continually cut and reassemble their own technological forms. Consumption has become, in other words, a self-generating machine whose only "function" is to reproduce an increasing surplus of its own technological style, its own simulacral technology—a surplus value whose only end is more consumption, more sales. In this technology of technological reproduction, the functional, scientific pretensions of the avant-garde's machine aesthetic give way to a technological aesthetic increasingly deferred from functionality or instrumentality. The avant-garde technē, in other words, comes to be replaced by a new, "postmodern" technē: high technē.

4.
Within the Space of High Tech

Much like industrial technology before it, "high technology," in its various forms, has been the subject of innumerable claims and charges. Nothing has been more common than to hear a given development in high tech celebrated for its ability to improve human life or condemned for its potential to destroy human values. On the one hand, for example, the "information superhighway" will supposedly increase our access to knowledge, expanding our creative potential and making us better able to control our lives, while at the same time providing a wealth of services and entertainment for our enjoyment. On the other hand, the "information superhighway" threatens to alienate us from human contact and community, while subjecting us to increasing degrees of surveillance, not to mention dividing us into the "information rich" and the "information poor." Whatever the merits of these arguments, there is little in them that is specific to high tech; indeed, the basic terms of this debate are the same as those that have been advanced since the very beginnings of modernity: technology is cast as either liberatory or alienating, utopian or dystopian, according to whether it is seen as an aid to humanity or not. Despite the many changes that have been associated with it, then, high tech continues to be conceived in the same terms that have defined technology throughout modernity. It continues, in other words, to be defined in terms of instrumentality, either as an instrument or tool that serves humanity or as out of control and therefore dangerous, dystopian. Thus, although commentators often seek to distinguish high technologies from modern technologies by contrasting electronic to mechanical, digital to analog, the diffusion of power to its centralization, at the same time they continue to view high tech in terms of modern technology: as merely a further extension of an instrumental or technological rationalization, as simply a more efficient, more "technological" version of modern technology.

Yet, even as these debates continue to figure technology in terms of instrumentality, it is evident that the very notion of high tech involves a profound change in how technology is conceived. In important respects, high tech can no longer be defined *simply* in terms of the modern, instrumental conception of technology. This is not to say, of course, that with the inception of high tech, technology has suddenly become entirely noninstrumental. Especially when considered at the level of a specific technology—a computer, a video camera, a magnetic resonance imaging scanner, and so on—high tech remains thoroughly instrumental, a means of achieving particular ends. It is only when high tech is considered as a general phenomenon that a shift in the conception of technology begins to become apparent. Understood in this more general sense, high tech comes to be seen as something that is too large, too complex, too uncontrollable, to be considered a means or tool to some rational, predictable end, whether that end involves a "humanist rationality" or "capitalist rationalization." It comes to be seen as a process in its own right, which functions according to its own logic, its own "aesthetic" dictates. Bernard Stiegler has, in fact, argued that "the *autonomy* of the machine" must be seen apart from the "laws of a universal human intentionality that no longer has a purchase here":

> Accounting for the technical dynamic non-anthropologically, by means of the concept of "process," means refusing to consider the technical object as a utensil, a means, but rather defining it "in itself." A utensil is characterized by its inertia. . . . The industrial technical object is not inert. It harbors a *genetic* logic that belongs to itself alone, and that is its "mode of existence." It is not the result of human activity, nor is it a human disposition.[1]

The idea of high tech, then, involves not only a shift in the conception of technology, and of the aesthetic, but also a shift in the very definition of humanity. Modernity has, after all, tended to define the human subject in terms of its presumed mastery of technology and the world. From a modernist perspective, a technology that can no longer be seen simply as an instrument under human control becomes a monstrous, dystopian threat to humanity's sovereignty. Indeed, many contemporary critics tend to view the increasingly complex and fluid processes of the techno-cultural world as just such a threat. Yet, we can also see the beginnings of a different view, in which the complexity and autonomy of high tech, of a techno-cultural world, is not simply seen as a dangerous other. Here, however, humanity can no longer be defined in terms of mastery and control, as "*the* human subject."

If we are to understand this shift in our relationship to technology, however, we must understand how the change in our view of technology came about. This shift cannot be seen simply as a radical break, as the replacement of one notion of technology by an entirely unrelated one. The techno-logic of high tech does not simply arise out of the blue, but develops out of the modern conception of technology itself. Indeed, it is as a result of its own attempts to extend itself, to replicate itself, that modern instrumental rationality brings about its own mutation, transforming itself into high tech. Like all mutations, then, this one comes about through a process of replication, of reproduction. It is, in other words, through what Benjamin calls "technological reproducibility" that the mutation from an instrumental notion of technology to a noninstrumental techno-logic occurs.

Technological reproducibility is, after all, generally taken to begin when an instrumental rationalization is extended to the realms of artistic and cultural production, as exemplified in aesthetic modernism generally, and particularly in the ideas of the modernist avant-gardes. Yet, in the modernist attempt to rationalize art and culture, the instrumental conception of technology itself seems to change, to become more a matter of "aesthetics," of culture. The techno-logic of high tech is based precisely on an aesthetic-cultural logic, which is also to say, on the logic of technological reproducibility. In high tech, technology comes to be defined in terms "of reproduction rather than of production."[2] Thus, the technological reproduction, alteration, and reassembly of signifying elements in high tech becomes less a means to an end than an artistic-cultural process that has become an end in itself. In high tech, in other words, the very function of technology has become more a matter of technological reproducibility than of an instrumental rationality.

If, however, we are to understand how this mutation in the conception of technology comes about, it may be helpful to compare this techno-logic of reproducibility to that noninstrumental "essence" of technology which Heidegger sees as ongoing within modern technology's instrumental Enframing of the world. At first glance, this comparison may seem far-fetched. Heidegger, after all, does not find this noninstrumental or "nontechnological" essence in contemporary technological reproducibility. Indeed, there is evidence that Heidegger sees technological reproduction as simply a further instance of an instrumental or technological Enframing, as when he notes that "hearing and seeing are perishing through radio and film under the rule of technology."[3] Instead, Heidegger sees this "nontechnological" essence of technology in the ancient Greek notion of *technē*. For Heidegger, technē exemplifies a conception of technology that is more closely related to artistic production *(poiesis)* than to modern technological production.[4] Heidegger does not, however, simply fall into the modern opposition between the technological and the aesthetic. For him, this more artistic or poetic production is not only "nontechnological," it is also "nonaesthetic"; it neither "enframes" the world in instrumental terms nor simply categorizes its "objects" as aesthetic. Rather, as in the production of art, it "brings forth" the elements of the world, allows them to come to representation, by "unsecuring" them from fixed meanings or values—including both use-value and aesthetic value. Whereas modern technology tends to "regulate" and "secure" objects in terms of their potential use to humanity, and aesthetics tends to fix objects in terms of an "eternal" aesthetic value, this more poetic mode of representation or production continually breaks things free of a stable context or fixed representation, representing them instead as part of an ongoing process or movement—as part of a continual process of unsecuring.

Thus, Heidegger's "essence of technology" can be read, as Samuel Weber has suggested, as an unsettling process or movement, which, rather than simply Enframing the world, setting it in place, also "dismantles" or "unsecures" it.[5] And, whatever Heidegger's objections to such a comparison might have been, it is worth noting the degree to which this process of "unsecuring" resembles the logic of technological reproducibility. Not only do both serve to break signifying elements free of their previous contexts, but they are both related to the modern, instrumental conception of technology in similar ways. First, modern technology's ordering or Enframing of the world in instrumental terms would be impossible without an unsecuring capable of "dismantling" a more holistic, mythic, or "enchanted" view of the world. According to Benjamin, technological reproducibility works in a similar way to "dismantle" the "enchantment" or "aura" of the aesthetic realm, enabling art to be represented in functional or instrumental terms. In both cases, the world or art object can only be represented in instrumental terms once its sense of "living" wholeness—its "spirit" or "aura"—has been dismantled, broken into elements freed of "inherent" presence or meaning, into empty signifiers. Second, the modern ordering of the world in technological or instrumental terms is itself a reaction to the ongoing process of unsecuring. Thus, the more elements are unsecured from a fixed meaning or context, the more desperate is the effort to regulate and control them, to resecure them in the terms of human use and control. With the rise of technological reproducibility, this unsecuring increases exponentially, as do the efforts to control it. Yet, because the Enframing involved in modern technology occurs by means of a process of unsecuring (which in high tech is increasingly carried out through technological reproducibility), every effort at regulation and control also involves a further unsecuring, which then requires further efforts to regulate and secure it. In a high-tech world, this effort to maintain instrumental control over those elements or signifiers unsecured by technological reproducibility only leads to a further proliferation of technological reproductions, with the pace of the cycle constantly accelerating.

Caught in this vicious circle, modern instrumental rationality begins to turn on itself, and in the process, to turn itself into something else. Carried to its "rational" extreme, the extension of an instrumental Enframing brings about its own mutation, its own unsecuring, allowing the unsettling techno-logic of reproducibility within it to come to the fore, no longer subordinated to an instrumental rationale. Although this mutation begins in modernist art's abstraction of form from function, it only comes to be seen as a mutation when the rationalization of the cultural world reaches a stage where its technological reproductions become so numerous and the interactions among them so complex that they are no longer subject to rational prediction and control. At this stage, technological reproducibility—taken in a general sense—can no longer be seen simply as instrumental, as a means to an end. Instead, the reproduction, alteration, and reassembly of elements removed from their previous contexts becomes an end in itself. Stripped of both aura and instrumentality, these elements become "purely" stylistic or "aesthetic"—empty signifiers that can be recombined in virtually any way.[6]

In other words, technological reproducibility becomes, like that artistic production and representation that Heidegger sees in the Greek *technē*, a form of *poiesis*: an ongoing process of representation or producing that both depends on a continual unsecuring and continually unsecures its own representations. Here, as Heidegger claims of the Greek *technē*, technology and art begin to converge—not in the form of the living or aestheticized technology that haunts so much of aesthetic modernism, nor in the form of the entirely functional, technologized aesthetic imagined by the modernist avant-gardes, but in an unregulated, autonomous process of technological reproducibility. Technological reproduction becomes increasingly similar to artistic production, not only because artistic production comes to depend more and more on technological reproducibility, but because the process or logic of technological reproducibility itself comes increasingly to be seen as "aesthetic," as a matter of style. If, in the era of artistic modernism, technological reproducibility was associated with an aesthetic of collage and montage, governed by the notion of functionality, in an era of high tech, technological reproducibility has become part and parcel of an aesthetic of pastiche, whose function or purpose cannot be determined in advance, but only in context.

High tech is defined not only by the techno-logic of reproducibility, but also by this aesthetic of pastiche. In fact, as the function of high-tech devices has become increasingly a matter of reproduction, it has also become a matter of reproducing and multiplying cultural styles. Television sets and VCRs, for example, may have an instrumental purpose, but that purpose is itself a matter of reproducing and rearranging cultural artifacts and cultural styles: of producing a cultural and aesthetic pastiche that can no longer be seen simply in terms of functionality or instrumentality. Technology, in short, comes to be seen as an "aesthetic" process, although this "aesthetic" lacks the wholeness of the aura. Thus it is that the conception of technology in high tech can be described as a kind of high technē.

Although modernist aesthetics seemed unable to acknowledge that its reproduction of technological style served to separate technological form from technological function, this separation is often readily apparent in high-tech style. In this context, it is worth noting that "high tech" has been used not only to describe "advanced" technologies, but also as a term of architecture and design, where it was coined from "a play on the words *high-style* and *technology*."[7] Used in this sense, "high-tech style" has been defined by its "imitation" of functionalism, or more precisely, by its imitation of the functional *style* of factories, warehouses, and industrial design generally.[8] Thus, for example, "high-tech" architecture has been described by its borrowing of technological elements, of a technological form: "when a residence sports open-web steel joints, corrugated aluminum siding, roll-up loading-dock doors, and/or steel mezzanines—components more commonly used in the construction of factories, warehouses, and schools—it's called high-tech."[9] This reproduction and remotivation of technological form is particularly evident in one of the most noted examples of the "high-tech school" of architecture, the Centre Pompidou in Paris. By separately color coding its structural, mechanical, and service technologies and making them part of the building's exterior, the Centre Pompidou highlights its technological style. Indeed, it takes on the look, as Kenneth Frampton has noted, of "the oil refinery whose technology it attempts to emulate."[10] Yet, in this imitation or reproduction of technology, technological form has clearly been separated from function; the building obviously does not function like an oil refinery; it merely abstracts and reproduces the "look" of a refinery, its technological style.

High-tech style, then, acknowledges this separation of technological form from function in a way that the modernist avant-gardes, with their belief in "functional form," never could. Thus removed or unsecured from their previous, functional context, these reproduced technological forms can be "reutilized" as stylistic or aesthetic elements. Through this process of unsecuring—that is, through technological reproduction—the conception of technology is itself unsecured; no longer chained to a notion of functionality or instrumentality, it comes instead to be defined in terms of a technological style, a high-tech "aesthetic."

This emphasis on technological style is not, however, limited simply to "high-tech" architecture and design, but defines high technology generally. In high tech, as the following definition shows, not merely the design but the very function of technology comes to be defined in stylistic or aesthetic terms—as *state-of-the-art*: "'High tech,' as the term is often used in shorthand, refers to the application of science to products that are at the state of the art in terms of their function and design."[11] Although it is common enough to hear high technology described as "state-of-the-art," the aesthetic metaphor buried within this term is generally ignored, so that state-of-the-art technology is generally taken to mean no more than advanced or highly developed technology. Yet, to be at "the state of the art" implies participating in a process or movement of constant technological change. It is to remain at "the edge of the envelope," to be on "the cutting edge," "the leading wave," or any of the other contemporary metaphors of vanguardism. And indeed, to be at the state of the art is to be, in a way, *en avant.*

In the age of high tech, vanguardism has become a corporate strategy, through which the newest versions of hardware and software are marketed and sold. Indeed, corporate success is itself explained as a result of being on the technological "leading edge," at the "state of the art." Thus, for example, corporations such as Microsoft and Intel have continually justified their successes as resulting not from monopolistic practices, but from the fact that "our technology was better"—which is to say, more advanced, more *avant-garde.* This rhetoric of corporate vanguardism is, of course, merely a justification for corporate greed, in much the same way that earlier, industrial and colonial, forms of exploitation were figured as "pioneering" or "trailblazing."

In fact, corporate success in technology seems to have less to do with being in the technological vanguard than, as it always has, with controlling channels of distribution, of access. Indeed, it is precisely because the process of technological change is notoriously difficult to predict or manage that this type of control—which can only take place after the fact, not before it—is necessary. As state-of-the-art technology, high tech is a "relatively autonomous" process that is always in advance of attempts to control it or profit from it. Human beings, or corporations for that matter, can at best hope to participate in this process, to "cooperate" or "interact" with it. It is, moreover, an "aesthetic" process, in which both the design and the function of technology are determined by neither instrumental nor capitalist concerns, but by the shifting logic of aesthetic or cultural style.

The notion of a high-tech style or aesthetic is therefore less a matter of a particular style than of a more general process of stylistic reproduction, alteration, and pastiche. A wide range of objects and designs have been seen as having a "high-tech style," from minimalist, "functional" interior designs to the baroque complexity of microprocessors to the techno-symbolic exteriors of high-tech architecture. In many cases, in fact, the appellation of "high tech" has been extended to items that seem to involve only the vaguest reference to anything technological. When we find "high-tech" used as an adjective for everything from track lighting to track shoes, it is obvious that the term does not refer to any particular notion of technological style, but to a much broader sense of "cutting-edge" aesthetics, of stylishness. High tech, in other words, refers to that stylistic "edge" or "wave," to that technological process or movement, in which elements are constantly abstracted, reproduced, and unsecured from their previous context: videotaped, digitized, altered, reassembled, recontextualized.

This process, however, has two basic aspects or tendencies, which can be seen in the design of virtually any piece of high-tech electronic equipment, where a seemingly incomprehensible interior complexity is linked to a severely minimalist exterior, both of which are seen as exemplifying a "high-tech style." Thus, while high-tech style seems at times to manifest itself in a tendency toward formal minimalism, at the same time it is associated with formal complexity. Although, at the level of the individual high-tech device, these two aspects of high-tech style seem to have little relation to one another, at a more general level, they are very much related, and explain a good deal about how the conception of technology changes from modernism to high tech.

At first glance, the minimalist exterior design of many high-tech components would appear to be little more than an extrapolation of the modernist aesthetic of functional form, which is itself an extension of modernity's drive toward an increasingly technologized or rationalized world. Just as factory mass production was based on the rationalization and standardization of parts, every element of the high-tech component seems to have been abstracted and reduced to its minimal, most rationalized form. Even the functional switches and control knobs that were so prominent in earlier, modernist-influenced designs have been placed behind hidden panels or replaced with flush-mounted touch controls so as not to disturb the abstract simplicity of the high-tech design. Indeed, the "ideal type" of the high-tech exterior would seem to be a "black box," stripped of all ornament, even color, and reduced to its most basic geometric form. Here, just as in modernist design, both the device's formal abstraction and its lack of color seem designed to convey a sense of functionality.

Yet, the "black-box" design of the high-tech component's exterior actually has little to do with its interior functioning; its minimalist, functional look is almost entirely symbolic, serving largely as a signifier for the complex, "state-of-the-art" technology inside it. In this connection, it is worth noting that the term *black box* has been used in electronic and computer design to designate any component whose interior functioning is either unknown or need not be considered.[12] This is in fact how most people view high-tech hardware: as a "black-box" technology, in which a minimalist exterior design serves as an aesthetic or stylistic representation of a "state-of-the-art" interior technology whose functioning remains unknown, seemingly beyond our comprehension.

As this minimalist, "black-box" tendency is pushed further and further, however, technological form is reduced to such a minimum that it begins to move "beyond black" and toward a kind of invisibility or transparency. This tendency toward technological invisibility can be seen, for example, in the trend toward the miniaturization of electronic and other technological components. Advances in semiconductor fabrication techniques have already allowed transistors and other semiconductors to be reduced to the microscopic level, and the development of nanotechnologies promises to extend this miniaturization even further—from the microscopic to the molecular level. The results of this trend toward miniaturization can be seen across a wide range of products, from cellular flip-phones and pagers, to laptop computers and personal digital assistants, to personal audio and portable video equipment. In each of these examples, an attempt has been made to reduce technological form to the minimum necessary for interaction. Research scientists, however, continue to work toward the ideal of a completely invisible technology: the dream of a computer and communications technology that would be "ubiquitous" and yet entirely "transparent." Such a dream is, of course, merely the computer-age version of André Bazin's dream of a "total cinema," in which the technology of cinema would be effaced in favor of a "complete illusion" of reality.[13] Perhaps the best example of this dream of an invisible yet ubiquitous technology can be found in the notion of virtual reality (VR).

The very idea of virtual reality implies the transparency of the technologies that produce it. The function of virtual-reality technologies is to allow users to be "immersed" in the "reality" that they present, to make that "reality" as fully present as possible. In order to achieve this experience of "immersion," virtual-reality technologies must efface their own technological form, make that form transparent or invisible to their users. The nearer that VR technology comes to achieving this transparency, the nearer it comes to being a perfectly functional form—a form that follows its function so closely as to efface itself completely. A perfect VR technology, in other words, would be a *purely* functional technology.

Yet, at the same time, the example of VR technology demonstrates the impossibility of this dream of an invisible, purely functional technological form. VR technology may be able to efface its "hardware"; it may be able to "minimize" its technological form to the point of invisibility, but its technological form does not, for all that, simply disappear. Rather, it undergoes a mutation: instead of referring to technological hardware, technological form becomes a matter of *software*—a matter of information, of data. It becomes, in other words, a matter of *simulation*; the simulated, virtual reality of VR *is* its technological form.

The example of VR demonstrates, then, that the minimalist tendency of high-tech aesthetics is not limited to exterior design, nor even to technological hardware itself. In VR, "reality" too is "formalized," "aestheticized": it is subjected to an aesthetic that abstracts and reduces it to its minimal elements, to the status of (invisible) bits of information. Here, "reality" has itself become a technological form, technologically reproduced and reproducible. "Reality," in short, is *digitized.*

This digitization of "reality" is the logical extension of the minimalist, functionalist aesthetic that high-tech style borrows from modernism. As such, it is also an extension of the technological rationalization of the world, through which "reality" is abstracted and reduced into ever more minimal—and potentially more controllable—elements. Paradoxically, however, this ongoing rationalization leads to a proliferation of the very elements it strives to bring under control. As this process of rationalization divides the world into ever smaller fragments, into "bits" of information, more and more of these "bits" are produced—and reproduced. Information proliferates, and, as it does, the "world" of information becomes increasingly complex.

Thus, the minimalist, "black-box" aesthetic of high tech is very much related to the aesthetic complexity that is also part of the high-tech style or aesthetic. The tendency to abstract and rationalize "reality" into ever more minimal forms or elements leads to a profusion of forms, elements, and information, and thus to an increasing complexity or "density" of style, which is also associated with high tech. Just as the minimalist tendency of the high-tech aesthetic can be seen as figured in the form of the high-tech exterior, this complexity of information or style is represented by the interior of the high-tech device: in the minute complexity of its circuitry, in the density of information stored there. Noting that microprocessors are often "represented as grid-like webs," Thierry Chaput describes their aesthetics as precisely "a matter of complexity, of the endless proliferation of their networks and the extent of their combinatories."[14] For Chaput, in fact, this complexity is such that microchips come to have "a secret aesthetic of their own," because it is no longer possible for any one person to understand them (p. 184).

Yet, if this complexity defines the aesthetics of the microchip, it also serves to define the look or style of contemporary techno-culture more generally, as Chaput notes when he observes that the microchip aesthetic "puts us in mind of a town like Los Angeles" (p. 185). Anyone familiar with the Los Angeles metropolitan area will of course understand this comparison: living in the city's vast grid of neighborhoods, streets, and buildings, not to mention its sheer physical and cultural density, does suggest what it must be like to live within an immense, and immensely complex, printed circuit. Whereas the space of the modernist city, exemplified in *Metropolis,* was represented in terms of abstract or minimalist forms still under the control of an instrumental rationality, the technological space of contemporary metropolitan areas such as Los Angeles has come to be seen as too mixed, too complex, to be figured in rational terms. If *Metropolis* suggested the danger of a controlling technological rationality leading to a chaotic, destructive return of the repressed, the complex, techno-cultural pastiche that is Los Angeles represents precisely the return of what modernity attempted to repress, which is undoubtedly why Los Angeles evokes in many the fear that "things have gotten out of control."

As the example of Los Angeles demonstrates, then, in an age of high tech, the complexity of the high-tech interior—of microchips, circuit boards, and computers—has become a common way of representing a sense of the increasing technological "complexification" of the world. Yet, this linkage of the complexity of high-tech microcircuitry and that of the technological city is more than merely metaphoric; for, as microchips become increasingly complex, miniaturized, dense, their functioning more incomprehensible, their form more "secret," the world comes to be seen as similarly complex and incomprehensible, as indistinguishable from technology—a technology that is no longer simply the instrument of human knowledge and control.

Among others, Jean-François Lyotard has linked the "process of complexification" of the world not only to the development of new technologies and of techno-science generally, but to the notion of postmodernity.[15] One could easily argue, in fact, that a sense of complexity has been a defining factor—and perhaps *the* defining factor—in every conception of postmodernity. Yet, as the vogue of theorizing postmodernity recedes, it becomes more and more apparent that this sense of a postmodern complexity was always a matter of a technological—or rather, techno-cultural—complexity. Indeed, if "postmodernity" has any meaning at this point, it is perhaps simply as a designation for this sense of being immersed in an immensely complex techno-cultural space.

It is precisely this sense of the complexity of "postmodern," technological space that Fredric Jameson, for example, finds in his analysis of postmodernity. Jameson, too, connects postmodernity to Los Angeles, but for him the attempt to figure this "new postmodern space" is "presently best observed in a whole mode of contemporary entertainment literature" that he dubs "high-tech paranoia."[16] In this literature, as he notes with some sarcasm, "the circuits and networks of some putative global computer hookup are narratively mobilized by labyrinthine conspiracies of autonomous but deadly . . . information agencies in a complexity often beyond the capacity of the normal reading mind" (p. 38). According to Jameson, then, not only do such narratives link the technological complexity of high-tech circuitry and computer networks to that of the larger techno-cultural world, but the complexity of these narrative worlds verges on being beyond comprehension.

Jameson suggests that "such narratives . . . have only recently crystallized in a new type of science fiction, called *cyberpunk,*" and cites the work of William Gibson as exemplary of this tendency.[17] Indeed, Gibson's work—not only in its portrayal of a complex techno-cultural world, but also in its sheer narrative density—seems almost designed to convey the sense of living within an inescapably complex techno-cultural space. In works such as his so-called "sprawl" series of novels and stories—which includes the *Neuromancer* trilogy and several stories from *Burning Chrome*[18]—Gibson fills his narratives with a dense mixture of technology, brand names, corporate intrigue, and details of techno-pop culture ranging from VR soap operas to computer hacking to body modification, from stylized underclass gangs such as the Gothicks and the Casuals to technological dropouts such as the Lo Teks and technological terrorists such as the Panther Moderns. This wealth of narrative and cultural detail seems rather appropriate to the complex techno-cultural world that Gibson attempts to portray. Readers at times report feeling lost and overwhelmed by this profusion of details and by the vague hints of plots that are beyond human understanding, much as the characters in Gibson's science-fictional world find themselves caught within complex techno-cultural spaces such as the huge urban sprawls from which the series takes its name.

The dense urban space of "the sprawl" is constructed as much from pop and street culture as from high technology per se; elements of both technology and culture are reproduced, sampled, and mixed in a bricolage of techno-cultural images, signs, and data. Indeed, in the world of "the sprawl," this complex mix of technologically reproduced elements has become so pervasive that it comes to seem inescapable. Gibson's sprawl is in fact a space that seems to have no "outside"; one cannot step beyond it to a more authentic or "real" vantage point, view it from a more comfortable—or critical—distance. In the world of "the sprawl," even the sky, as *Neuromancer*'s now-famous, and often-cited, opening line suggests, conveys this sense of an inescapable techno-cultural space: "The sky above the port was the color of television, tuned to a dead channel."[19] Here, "reality" itself seems to have changed, mutated, to have become, like television, a complex grid of simulations, of data.

The dense, mutated, urban space of Gibson's "sprawl" is in many ways merely the logical extension of contemporary techno-cultural space, of the contemporary urban environment. Gibson has himself noted that in his science fiction he is "just trying to make sense of contemporary reality":[20]

> My SF [science fiction] *is* realistic in that all along I've been writing about what I see around me. I'm reacting to my impressions of the world. My fiction amplifies and distorts these impressions . . . but I try to present the way I actually perceive the world, at least certain glimpses of it. I'll be sitting in the Dallas/Ft. Worth Airport looking out my window and thinking, "What is this landscape, anyway?" . . . You know you're in a very strange place, but you're also aware this weirdness is just your world.[21]

Yet, if Gibson's sprawl world of the future is based on the contemporary urban environment, his comments also suggest that the present-day world is *already* "strange," "weird," mutated—which is to say it is already science-fictional. As Gibson himself observes, "the world we live in is so hopelessly weird and complex that in order to come to terms with it, you need the tools that science fiction develops."[22] Fellow "cyberpunk" Bruce Sterling makes a similar observation about the contemporary world when he notes that the cyberpunk authors were the first generation of science-fiction writers to grow up "in a truly science-fictional world."[23]

The "cyberpunk" suggestion that we are *already* living in a "science-fictional world" is simply another way of describing the sense that the world—and our relation to it—has changed, mutated. This "science-fictional world," in which we are "already living," is precisely the immensely complex, simulated, techno-cultural space of "postmodernity"; for, as both Gibson and Sterling make clear, the "science-fictional" space of Gibson's "sprawl" differs only in degree from the "postmodern" space of contemporary cities such as Los Angeles: in both cases, the city comes to be seen as an "interior" space, a space that we live *in,* rather than as an "external" world. This sense of being *within* a "postmodern" or "science-fictional" space can perhaps serve to explain why the figure of the postmodern city is so often conflated with the microchip, with the interior space of the computer or other high-tech devices, for we are already living "inside" technology, within a techno-cultural space or world. It no longer makes sense to think of the city, of the world even, as "outside" of technology.

The conflation of urban space and the space of the high-tech interior is also apparent in Gibson's now-famous description of future cyberspace, the matrix. Just as the dense, mutated space of Gibson's "sprawl" is in many ways merely the logical extension of contemporary, "postmodern" urban space, the matrix is itself an extrapolation of the "interior" space of the present-day computer or computer network:

> The matrix has its roots in primitive arcade games, in early graphics programs and military experimentation with cranial jacks. Cyberspace. A consensual hallucination experienced daily by billions of legitimate operators, in every nation, by children being taught mathematical concepts. . . . A graphic representation of data abstracted from the banks of every computer in the human system. Unthinkable complexity. Lines of light ranged in the nonspace of the mind, clusters and constellations of data. Like city lights, receding . . . [24]

In using the figure of a vast city receding into the distance to describe "the matrix," Gibson condenses the two most popular figurations of postmodern, techno-cultural space: the city and the computer (or computer network). Within this space, the "external" space of the city is mathematicized, digitized, transformed into a space "inside" of the computer, while the "interior" of the computer (or computer network) is given "graphic representation" in the form of the buildings, thoroughfares, and lights of a virtual city. In the matrix, then, the analogy between the urban grid and the circuitry of the microchip, between the space of the city and the interior space of the computer, is no longer metaphoric at all. The matrix is in fact simply a representation of the dense, already simulated urban space of the "sprawl": in the matrix, urban density becomes informational density, the city becomes a virtual space that no longer has an "external reality" but exists only within the computer, "inside" of technology. This is simulation taken to the point of virtuality, technology taken to the point where it disappears into its own interior, becoming a matter of information, of data. Thus, the city-space or grid of Gibson's matrix no longer simply *resembles* the grid of a television or video screen (like the sky of the "sprawl" world); it has, in a sense, already *become* video, already been transformed into a graphic representation of data.

Brooks Landon, in his book on contemporary science-fiction film, has in fact pointed to the crucial role that "contemporaneous film/TV/video" has played in the cyberpunk movement, noting that "it is hard to talk about cyberpunk writing *without* referencing it to contemporary film and TV."[25] Writing on Gibson's work, Landon argues:

> If you have seen *Blade Runner,* you have seen much of Gibson's future. If you have seen *Blade Runner* and *Tron,* you are already well grounded in almost all of the significant aspects of Gibson's semblance. . . . the special relationship of *[Neuromancer]* to *Blade Runner* and *Tron,* both of which were made before its publication, illustrates very well the reciprocal

> relationship that seems to characterize ties between cyberpunk and SF [science-fiction] media. (P. 122)

At the most obvious level, the dense, "retrofitted" techno-cultural space of Los Angeles in *Blade Runner* anticipates *Neuromancer*'s urban "sprawl," while the "video-game world"—depicted as the space inside a computer—of *Tron* provides an early analogue for Gibson's matrix. Landon is, of course, hardly alone in noting the influential status of the mise-en-scène in these two films. Vivian Sobchack, in her book on science-fiction films, has also pointed to *Blade Runner* and *Tron* as prototypes for the cinematic representation of "postmodern space," with *Blade Runner*'s "excessive scenography" representing a complexifying "inflation of space" and *Tron* enacting a "deflation" or flattening of space similar to that of a computer or video screen.[26] Yet Landon's point involves more than simply noting how cyberpunk writing has been influenced by the ideas and stylistic representations of particular science-fiction films and television programs. Rather, he points to how the contemporary media technologies of film, video, and computer imaging are *themselves* crucial models for the cyberpunk vision of a science-fictional world:

> the computer generated special effects "magic" of recent SF [science-fiction] film and the manic permutations and informational density of such music videos as Peter Gabriel's "Big Time," Cutting Crew's "One for the Mockingbird," and Tom Petty's "Jamming Me" become for the cyberpunk writers a key index to what everything will be like in the future—a time of designer drugs, designer genetics, designer surgery, designer prosthetics, even (courtesy of time travel) designer history. (P. 126)

Landon's analysis, here again, points to a certain affinity between cyberpunk and media—particularly music videos and contemporary science-fiction film. Certainly, cyberpunk writing—with its "quick-cut" narratives, its emphasis on startling and disjunctive images, and its "informational density"—has borrowed heavily from the stylistic techniques of music videos, as have such contemporary science-fiction films as *Hardware, Highlander,* and *Tank Girl.* Yet, for the cyberpunk writers, these stylistic borrowings are less important in themselves than inasmuch as they serve as a "key index" to the future. Indeed, it seems that for the cyberpunks, the style of certain music videos and the "special-effects 'magic'" of recent sci-fi films is *already* futuristic, science-fictional. What makes them seem science-fictional are the same qualities that have led music videos to be called "the first postmodern television": their "informational density" and their ability to reproduce, alter, and edit "reality," to design—and redesign—virtually anything, including the world itself. Indeed, as in many music videos, the science-fictional worlds imagined by cyberpunk are a dense, "postmodern" mix of cultural images, sounds, and data, all of which are subject to continual "unsecuring"—to reproduction, alteration, redesign, editing. In other words, what cyberpunk draws from music videos is precisely the sense of being within an "unsecured," "postmodern" world, a world of such "manic permutations and informational density" as to seem "strange," mutated, science-fictional.

Yet, if, as Landon suggests, cyberpunk is simply an extension of the already science-fictional, "postmodern" density and mutability of music videos, then the cyberpunk vision of the future, of a "science-fictional world," begins to sound very much like an "MTV world," an "MTV future." What Landon neglects to mention, however (even though his examples of music videos tend to confirm it), is that, much like the world of MTV, cyberpunk's science-fictional worlds have generally been produced from and oriented toward a white, heterosexual male perspective. At times, this connection between cyberpunk and MTV is quite blatant, as in such works as Norman Spinrad's *Little Heroes* and John Shirley's *Eclipse*, where the model for the cyberpunk hero is drawn directly from the image of the rebellious male punk rocker.[27] More commonly, however, the rebellious "punk" heroes of cyberpunk seem to derive from the largely young white male subculture of computer hacking, in which the sense of rebellion, desire for intensity, and romanticization of "street life" so apparent in young male drug and rock culture (including MTV) is transferred into the seemingly incongruous realm of computer and telecommunications technology.

In cyberpunk, however, the technological realm becomes precisely the space in which the masculine, outlaw ethos of drug culture and the street, MTV and punk, and computer hacking and phreaking converge. The densely urban environment of Gibson's sprawl is, for example, basically a more complex, more technologized version of contemporary life "on the street." Yet, if the technologized urban spaces of cyberpunk (as well as those in such science-fiction films as *Blade Runner, RoboCop,* and *Hardware*) are extrapolations from contemporary urban life, they also tend to extrapolate present-day urban problems such as overpopulation, urban decay, drugs, and crime. Thus, these spaces often resemble, as many commentators have noted, the seedy, urban world of the hard-boiled detective and film noir. Vivian Sobchack has in fact used the term *tech noir*—drawn, interestingly enough, from the name of a punk/heavy-metal club in *The Terminator*—to describe "the aesthetic of much recent SF."[28] This notion of "tech noir" serves, then, as a convenient name for the technological and aesthetic space where the worlds of high tech, noir, and MTV converge.

Just as the masculine codes that govern the actions of the hard-boiled noir detective can only be understood in the context of the noir world, the masculine-oriented conventions of cyberpunk can only be understood if they are considered in terms of the "tech-noir" environment that surrounds its male heroes. Cultural critic Claudia Springer is among those commentators who have noted the extent to which the technoscapes of cyberpunk mirror the mise-en-scène of film noir: "Cyberpunk's dark, bleak surroundings and its convoluted plot twists that often involve treachery and betrayal are derived from the cynical world of film noir."[29] As Springer suggests, then, the tech-noir world of cyberpunk, much like that of film noir, has generally been represented as a "dark" and dangerous urban space whose sheer size and density make it difficult, if not impossible, to discern the motives and forces at work within it. Indeed, the tech-noir city is in many ways an *amplified* version of the film-noir city: larger, more dense, and faster-paced. Cyberpunk and cyberpunk-related films often seem to suggest, in fact, that with this "amplification" in size, density, and speed comes a similar amplification of the problems and fears associated with the film-noir city. Gibson, for example, describes an area called "Night City" as "a deranged experiment in Social Darwinism, designed by a bored researcher who kept one thumb permanently on the fast-forward button."[30] Similarly, *Blade Runner*'s Los Angeles and *RoboCop*'s Detroit are represented not only as "dark" and decaying, as dense in both population and detail, but also as spaces of "insecurity," where corruption, crime, and extreme violence have become so pervasive that they seem to threaten the very boundaries of social order.

Indeed, the threat posed by the tech-noir city seems to involve precisely the dissolution or unsecuring of boundaries. Often, this threat is cast in terms of fluidity and mixture, with the city figured as a kind of seething cauldron or mélange that threatens not only to overflow the boundaries of order, but to swallow up the cyberpunk protagonist who finds himself immersed within it. As Case, the protagonist of *Neuromancer*, puts it,

> Stop hustling and you sank without a trace, but move a little too swiftly and you'd break the fragile surface tension of the black market; either way, you were gone, with nothing left of you but some vague memory in the

> mind of a fixture like Ratz, though heart or lungs or kidneys might survive in the service of some stranger with New Yen for the clinic tanks. (P. 7)

What is at stake for Case, then, is not simply death, but the dissolution of his own boundaries, his own self, in the dense fluidity and accelerated processes of the tech-noir city.

Case's fears, as Klaus Theweleit has shown, are not unique. In *Male Fantasies,* his two-volume study of the psychology of the German Freikorps troops, Theweleit explores how these soldiers' fears are cast in terms of a "dark," threatening fluidity.[31] Whether this fluidity is figured in terms of raging floods or tides, or swamps and mire, or as boiling or seething, it threatens to overflow the boundaries of order and decency, to infiltrate and contaminate, to swallow and dissolve the identity of both the nation and the individual. This fear of a "dark," at times deceptive, but always threatening, fluidity is often associated with otherness, whether ethnic or racial, cultural or sexual. Theweleit, in fact, draws particular attention to the linkage of such fears to the fear of "erotic femininity." Faced with envelopment by these dark, female fluids, the soldier-male reacts with a "movement of stiffening, of closing himself off to form a 'discrete entity'" (p. 243). He hardens or "steels" himself, armors his body so as to protect himself, to secure his boundaries, against contamination and dissolution by this fluid otherness. A quite similar reaction is obvious in the "hard-boiled" detective, who, it is suggested, develops his tough, cynical "shell" to protect himself as he "plunges" into the "dark, seething waters" of the noir city, with all of its dangers and deceptions. Similarly, too, the dense mixture of crime, violence, and otherness that makes up the tech-noir city provokes a "hardening" of the male protagonist who finds himself immersed in it—whether psychologically, like Deckard in *Blade Runner* or the mercenary, Turner, in Gibson's *Count Zero,* or literally, as exemplified in the armored bodies of RoboCop and the Terminator.[32]

Given the high-tech protective armor of Robocop and the Terminator, one might be tempted to assume that, in this contrast between "hard" masculinity and "feminine fluidity," technology is—as it has so often been—masculine. Yet, if the threat posed by the dark, fluid density of the tech-noir city is clearly figured as feminine, as other—because, often, these urban spaces are also explicitly multiracial and multicultural—then it is worth remembering that the tech-noir city is itself technological. Indeed, Theweleit himself, drawing on the work of Deleuze and Guattari, has pointed to a connection between the fears of a fluid, "feminine" otherness and the fears associated with machinery.[33] Thus, technology is often represented as a "wave" that threatens to "overflow" the "security" of traditional boundaries, enveloping or "swamping" "humanity" in the process.

This sense of being enveloped by technology, immersed in an "insecure" technological space, is precisely what the tech-noir cities of cyberpunk and contemporary science-fiction films are designed to convey. Indeed, as Landon has argued, the cyberpunks' sense of already living in a science-fictional world is founded on the premise that, at least in highly technologized societies, technology—such as the media and computer technologies so prominent in music videos and recent science-fiction films—has become such an inseparable part of our everyday cultural life that we now feel ourselves surrounded by, immersed in, a new, techno-cultural environment:

> Cyberpunk is probably the first science fiction to take the cultural implications of technology completely seriously, to realize that velcro and video games have changed our world much more than has spaceflight. . . . Perhaps more than anyone writing today in any form or genre, the cyberpunks realize that electronic and medical technology now surrounds us, not as tools or toys, but as a new environment, an ecosystem that influences almost every aspect of our existence. (P. 123)

Although Landon's description of this technological change seems to rely on a technological determinism in which changes in technology dictate cultural effects, it also suggests the difficulty of distinguishing cause from effect, the technological from the cultural: Is a video game, for example, more a matter of technology or of culture? Are any of the various electronic and computer media that surround us every day—from television to videotapes to computers—more technological than cultural? If the cyberpunks no longer see technology simply in instrumental terms, as "tools or toys," but as a new environment, this is also to say that they see technology as increasingly indistinguishable from culture itself. Like technology and art, technology and culture have begun to merge. This is especially the case with media and computer technologies, which are designed largely to technologically reproduce cultural/aesthetic forms and data. People watching a film or a video, or playing a computer game, tend to experience technology only through the cultural images, sounds, and data that these technologies produce and convey. At the same time, culture itself, as represented through these technologies, becomes increasingly a matter of technologically reproducible information or data. Just as with virtual-reality technologies, then, as these media and computer technologies become increasingly miniaturized, complex, invisible, and ubiquitous, their technological form tends to disappear into cultural forms, to become part of techno-culture. Moreover, as ever more aspects of the world become subject to technological reproducibility, simulation, and digitization, the scope, density, and complexity of techno-culture also increase, until it comes to be seen as the very environment, the world, in which we live.

If, however, the techno-cultural environment of the tech-noir city is often figured as a "dark," dense, "feminine" fluidity into which the male protagonist must "plunge" or immerse himself, this figuration applies a fortiori to the even more fluid techno-cultural world of the matrix, of cyberspace. Nicola Nixon has in fact argued that the matrix, as represented in Gibson's trilogy, "is figured as feminine space," at once exhilarating and threatening to the male operators who enter it: "The console cowboys may 'jack in,' but they are constantly in danger of hitting ICE (Intrusion Countermeasures Electronics), a sort of metaphoric hymeneal membrane which can kill them if they don't successfully 'eat through it' . . . in order to 'penetrate' the data systems of such organizations as T-A (Tessier-Ashpool)."[34] Amplifying Nixon's point, Claudia Springer notes that "The word *matrix,* in fact, originates in the Latin *mater* (meaning mother and womb)," and goes on to suggest that cyberspace evokes the same uncanny feelings of "simultaneous attraction and dread" that are, according to Freud, evoked by the womb.[35] Along similar lines, Eva Cherniavsky has argued that in *Neuromancer,* cyberspace, constituted as it is between the poles of "originary plenitude" and "symbolic lack," "signifies . . . the fetishization of the maternal."[36] This connection between the matrix and the maternal is supported, as Nixon, Springer, and Cherniavsky all point out, by the fact that the matrix is also figured as a reproductive or generative space, in which technology—in the form of AIs—reaches a stage of such density and complexity that it comes to have a fluid and mysterious life of its own, no longer subject to "human" comprehension and control.[37]

The point at which this technological, other "birth" takes place becomes known, in the third volume of Gibson's trilogy, as "When It Changed," a change that, as Nixon observes, is figured as "the uncontrollable feminizing of the matrix, the uncheckable transformation of viral software into a feminine Other" (p. 227). The Finn, a character in *Count Zero,* in fact describes the matrix's technological life precisely in terms of an uncontrollable fluidity, femininity, otherness: "there's things out there.

Ghosts, voices. Why not? Oceans had mermaids, all that shit, and we had a sea of silicon, see?"[38] Yet, this otherness is not only represented in Gibson's work as feminine; it is also, quite explicitly, connected to "other cultures," to the "Third World." Thus, the AIs that come to life in the matrix appear in *Count Zero* in the form of Afro-Haitian loa, while in *Neuromancer,* the Neuromancer AI has Brazilian citizenship and is known through most of the novel only as Rio; when Neuromancer appears to Case, it is in the guise of a laughing "brown boy," barefoot on the beach.[39]

Although we might condemn—as, for example, Cherniavsky does—Gibson's appropriation and "exoticization" of other cultures (as well as his "fetishism" of the "feminine"), it is worth noting that Gibson also suggests that this "other life" is the result of a return, within technology itself, of precisely that which has been repressed by patriarchal, Western scientific-technological thought and culture. This return of the other is neatly summarized in *Count Zero* in the story of Wigan Ludgate, a cyberspace cowboy who "stormed off on an extended pass through the rather sparsely occupied sectors of the matrix representing those geographical areas which had once been known as the Third World."[40] Hacking the relatively unprotected computers of African banks, "the Wig" makes himself rich, although "incidentally bringing about the collapse of at least three governments and causing untold human suffering." Yet, as he attempts to enjoy his riches, the Wig is haunted by televised pictures of "the bloated bodies of dead Africans," until he becomes "mad," convinced that God can be glimpsed in ghostly forms that inhabit the matrix. That these ghosts in the matrix represent a return of the other (from the dead?) repressed and exploited by Western colonial and technocratic powers—represented here by the Wig's unthinking exploitation of the Third World's already "sparse" techno-space—is suggested not only by the fact that these "ghosts" appear elsewhere in *Count Zero* as Afro-Caribbean gods, but also by the fact that a penitent Wig begins to wear, as one character describes it, "all this African shit, beads and bones and everything."

What is, therefore, figured in the matrix is precisely the frightening but alluring prospect of *a technological other,* the specter of a technology that seems to have a life of its own—a ghostly, feminized, other life that is beyond the knowledge and control of patriarchal, Western science and technology.[41] As the space of this other, technological life, the matrix can only be seen as an environment, a world, unto itself, with its own autonomous, uncontrollable existence. Thus, as soon as the Finn, in the passage cited earlier, speaks of the ghosts and voices that populate the matrix, he immediately comments, "Sure, it's just a tailored hallucination we all agreed to have, cyberspace, but anybody who jacks in knows, fucking *knows* it's a whole universe" (p. 119).

If, however, the matrix is a world, a universe, with a wholeness, a life, an autonomy, of its own, it is exactly this wholeness that remains beyond the knowledge and control of the male protagonists who enter this world. As a space of "unthinkable complexity," the matrix is simply too vast, too dense, too complex to be comprehended in its entirety. There is, moreover, no external perspective from which it could be grasped as a whole; the matrix can only be viewed from within. Thus, there can be no map that would chart its overall space, no schematic diagram that would trace its complete circuitry. Any attempt to take in the matrix globally, as a whole world, can only yield a vague sense of it as a mutated, technological space (a cyberspace) beyond representation, a sense that is very much like the experience that Kant described as "the sublime." Yet, given the technological status of the matrix, a status that would have excluded it from Kant's notion of the sublime, it is perhaps more appropriate to speak of the matrix as the space of, to use Fredric Jameson's provocative phrase, a "*technological* sublime."[42]

For Jameson, this sense of a technological sublime is evoked by exactly the kind of complex, highly technological narrative worlds that are represented in Gibson's science fiction. In Jameson's view, however, representations such as Gibson's "of some immense communicational and computer network are themselves but a distorted figuration of something even deeper, namely the whole world system of present-day multinational capitalism" (p. 37). Now, at first glance, this diagnosis seems perfectly in keeping with the high-tech, decentered global network of multinational corporations described in Gibson's work. Yet, although both Gibson and Jameson recognize that this kind of high-tech world—both because of its complexity and because of our position within it—cannot be represented as a totality, Jameson still hopes to maintain a *knowledge* of this totality, of "the whole world system of multinational capitalism," that for him lies beneath, and indeed motivates, the profusion of techno-cultural images and data that make the high-tech world so complex. It is to this end that he proposes a "global cognitive mapping" through which individuals could understand their place within this complex global system, even while it remains unrepresentable, sublime.

Although Jameson, in employing the notion of cognitive mapping, attempts to find a means of "navigating" a complex, techno-cultural world from within it, the very phrase "global cognitive mapping" betrays his desire for a totalizing vision and knowledge (of the "whole world") that has traditionally been possible only from a distance, from a position outside of the world in question. Thus, even while Jameson observes that this kind of distance has been "abolished in the new space of postmodernism," he clings to a notion of distance as necessary to the very constitution of human subjects, for it is this "distance" that distinguishes them from objects and "lesser" life-forms, and from the world itself. Through the notion of "global cognitive mapping," then, Jameson attempts to maintain a "knowledge situation" based on this distinction of active human subjects from a passive object-world. From this perspective, a technological world that appears as beyond human comprehension and control, as autonomous, as an agency in its own right, could only be considered as dystopian, as a threat to "human" agency. Thus, Jameson struggles to maintain the utopian possibility of a human mastery over the increasingly dense, complicated, and "sublime" techno-cultural world of multinational capitalism.

It may well be that Jameson's description of this high-tech, postmodern world in terms of a high-theoretical concept like "the sublime" has distracted his many commentators from noting the rather close resemblance between his portrayal of this world and the representations of urban and technological space in film noir and cyberpunk. Yet, if the space of Jameson's postmodern world is sublime, it is also vast, disorienting, and dangerous to those who find themselves "immersed" within it. Indeed, this threat, much like that of noir and cyberpunk spaces, often seems to be cast in terms of a murky, threatening fluidity—in this case, the "flows" of multinational capital, which, respecting neither geographic nor individual boundaries, threaten to dissolve all critical distinctions, all history, into a huge, "unsecured," postmodern mélange. Individuals therefore find themselves "adrift" in a seething, turbid mix of images and data, unable to find their bearings, to locate themselves in relation to "the whole." Indeed, Jameson describes the schizophrenic "aesthetic" that he feels characterizes postmodernity precisely in terms of an "engulfing" or "overwhelming" fluidity, noting that "[the] present suddenly engulfs the subject with undescribable vividness, a materiality of perception properly overwhelming" (p. 27). Although Jameson notes that this fluid materiality need not be seen simply in "the negative terms of anxiety and loss of reality," but can also be viewed in "the positive terms of euphoria, a high, an intoxicatory or hallucinogenic intensity," his own anxiety and insecurity continually "resurface," continually figure the space of this "undescribable" postmodern fluidity in terms of a deceptive and threatening otherness.

The nature of this threat may be clearer if we consider that, in Jameson's "narrative," it is the critic who, much like the hard-boiled detective and the cyberspace cowboy, must attempt to "navigate" within the dense, murky waters of postmodern space without being overwhelmed, becoming lost, or sinking without a trace. These "waters," then, threaten to "swamp" the very "ground" of the critic's identity, to dissolve the boundary that distinguishes the "active, human subject" (of which the critic is, of course, the exemplary instance) from a passive object-world—and from the instrumental technology through which this subject knows and controls that world. Thus, although Jameson does not explicitly portray this threatening fluidity as feminine, he does position himself, as critic, on the side of an opposition that has historically cast an active, knowing human subject as masculine while regarding its "object"—nature, the world, and so on—as a wild, mysterious, and dark "feminine Other" to be penetrated, subdued, colonized, domesticated, or, in a word, "enlightened." And Jameson does, indeed, *explicitly* figure the postmodern or technological sublime as the space of a mysterious, "dimly perceivable" otherness. Noting that "The *other* of our society is . . . no longer Nature at all, as it was in precapitalist societies, but something else which we must now identify," Jameson goes on to identify it as "that enormous and threatening, yet only dimly perceivable, other reality of economic and social institutions" (p. 38). Thus, Jameson, like any noir detective, brings to light the "dark" and troubling "other reality" lurking beneath the deceptive fluidity of postmodern techno-culture; indeed, it seems that the main distinction between this "reality" and that of film noir is Jameson's substitution of "multinational capitalism" for "crime and corruption." The "enormity" of this postmodern space makes it unrepresentable (as a whole), other, sublime, and in fact brings it quite close to a notion of monstrosity that, as Jameson is surely well aware, is akin to the sublime. Indeed, it sometimes seems that what Jameson is describing is simply a postmodern version of Hobbes's Leviathan, its mechanistic aspects now computerized and decentralized to the point that it becomes fluid and amorphous, so that it, like technology, ceases to be visible in itself, and becomes simply the techno-cultural "reality" in which we live.

There is, of course, a remarkable similarity between Jameson's portrayal of the monstrous other that lurks within the postmodern or technological sublime and Gibson's science-fictional rendering of the complex and obscure other forces that come to exist within the matrix. In both cases, this technological other is presented as out of human control, as possessing an uncanny, Frankensteinian life—or at least agency—of its own. Indeed, these representations are quite similar to the monstrous, technological otherness of the False Maria in *Metropolis.* Yet, whereas Jameson—despite his admonitions to avoid moralizing judgments about postmodernity—ultimately sees this technological, other "life," like the False Maria, as a monstrous, dystopian threat, Gibson seems to view it not in the modernist terms of utopia or dystopia, but as a condition of possibility. The fact that the matrix cannot be comprehended as a whole, that it no longer seems subject to human knowledge and control, but rather, seems to have an autonomous agency of its own, does not seem to inspire in Gibson, as in Jameson, insecurity about humanity's own agency.

Unlike Jameson, Gibson does not seem to find it necessary to try to regain human control over the fluidity of the techno-cultural world, to "resecure" technology by putting it back "in its place" as a human instrument. Rather, Gibson seems to suggest that the appropriate response to this other world is interaction, not mastery. Among his characters, those who attempt to master the techno-cultural world inevitably find themselves undone by it. Those who survive do not, however, simply "mediate" between humanity and an autonomous technology; rather, they find ways to come to terms with the "insecurity" provoked by the techno-cultural world that surrounds them, to accept and live with the existence of other, nonhuman agencies. In doing so, they do not lose all sense of boundaries or agency; they do not become the slaves of a monstrous, mutant technological other, nor do they simply dissolve into the "dark," shifting fluidity of postmodern, multinational techno-culture.

Gibson's work suggests the possibility of a different relationship between human beings and the techno-cultural other—a relationship in which "humanity" would no longer be defined simply in terms of an active, "masculine" mastery over a "feminine," other world, but by its inseparable and multiple "connections" to that world. Rather than attempting, as Jameson does, to find new ways to maintain the space of human mastery, to secure the boundaries of the "human subject," this kind of connection takes the risk of remaining open to the fluidity and mutations of the techno-cultural world; for it is perhaps only by allowing ourselves to ride—and indeed, to live—within the unpredictable flows and currents of techno-culture that we can hope to learn, not how to control it, but how to hack its codes, to reroute the subroutines of its logic, in order to create new patterns of interaction, whose results we cannot yet foresee. After all, within the complex, global space of high-tech techno-culture, even small changes can yield sometimes startling mutations. This perhaps explains why this space so often seems to be figured as the space of a science-fictional, techno-cultural other: it is, after all, the space in which that unknown and uncontrollable mutation called "the future" comes into being. To move and live within that monstrous space is to live with the insecurity of that future, and with its promise.

5.
Technological Fetishism and the Techno-Cultural Unconscious

The "aestheticized," "state-of-the-art" quality of high tech may be—and has been—seen as a kind of fetishization of technology, of technological style. This fetishism of high tech is readily apparent in, for example, the pages of glossy techno-cultural magazines such as *Wired* and *Mondo 2000.* Indeed, *Wired* even includes a regular feature titled "Fetish," in which sleek new technological devices are treated as "sexy," aesthetic objects. Of course, the fetishism of technology is not unique to high tech. From the "stone age" to the "automobile age," technology often seems to have inspired a considerable degree of fetishism. In high tech, however, this fetishism often seems to have become more explicit, more a part of the very definition of technology itself.

Much of this sense of an increased fetishism of technology stems from high tech's shift away from a purely instrumental conception of technology. So long as technology was conceived as a matter of instrumentality, as a means for achieving practical ends, any noninstrumental value attached to it—such as an aesthetic or stylistic value—was necessarily auxiliary, supplemental. To the extent that this supplemental value came to be seen as having a value in its own right, it was viewed as a kind of fetishism. Thus, high tech's increased emphasis on the aesthetics or style of technology leads to the increasing sense of technological fetishism associated with it. In fact, the fetishism of technology seems to be inherent to very notion of high tech; in other words, high tech is, by definition, fetishistic.

This fetishism of technology in high tech obviously bears more than a passing resemblance to the fetishism of commodities analyzed by Marx. One could easily, for example, read the fetishizing of technology in magazines such as *Wired* and *Mondo 2000* as a textbook example of commodity fetishism: not only is a technological object viewed as having a unique value of its own, but, in the process, the conditions of its production and distribution are almost entirely effaced. Neither *Wired* nor *Mondo 2000* runs stories about the working conditions of the underpaid laborers in overseas semiconductor fabrication plants. Nor does either give much consideration to the reasons why high tech has been, and generally continues to be, the exclusive province of upper-middle-class white males. Indeed, these magazines often seem to assume that access to the highest levels of high tech is simply a matter of being smart enough, or hip enough, to want it. Thus, these magazines—and, presumably, many of their readers—tend to see high tech as having its own inherent value, an aesthetic or stylistic value, a certain "sexiness," that seems to reside in technology, just as, in Marx's analysis, it does in commodities.

Yet, the fetishism of technology in these magazines at times extends beyond simply the fetishism of particular high-tech objects; indeed, the very idea of high technology is itself fetishized, treated as having a value in itself. Thus endowed with immanent value, high tech tends to be seen less as a means or tool for human use than as something autonomous of human control. Generally, in fact, it is represented as a kind of autonomous movement or force, with its own complex logic and its own inscrutable ends. As such, high tech often comes to be seen as having a mysterious "life," or at least an agency, of its own.

The notion of an object endowed with a mysterious value, power, or life is, of course, the traditional definition of fetishism, a definition that Marx drew upon in formulating his idea of commodity fetishism. In fact, one of Marx's illustrations for the notion of fetishism is his famous metaphor of a table come to life, which,

> so soon as it steps forth as a commodity . . . is changed into something transcendent. It not only stands with its feet on the ground, but, in relation to all other commodities, it stands on its head, and evolves out of its wooden brain grotesque ideas, far more wonderful than if it were to start dancing of its own accord.[1]

From Marx's perspective, the value or life attributed to commodities is a kind of mystification that treats an object not in terms of the human needs it fulfills—its use-value—but as an independent or autonomous entity. It is therefore tempting to view the idea that high tech has a force, an agency, or a "life" of its own in similar terms: as a mystification, as a species of commodity fetishism in which the conditions of high tech's production and distribution are ignored.

Yet, despite its value in pointing to the effacement of technology's conditions of production and distribution, Marx's notion of fetishism is based on a highly Eurocentric view of the relationship between humanity, technology, and the world. It assumes, first of all, a sharp distinction between human subjects and an object-world that is viewed in terms of its potential use-value to humanity. Similarly, technology is seen as an instrument or tool for human use. Thus, any object or technology that is seen as having a value, much less a life or agency, beyond its instrumental value to humanity is necessarily viewed as fetishized. This instrumental view of the world is presented as rational, scientific, and "modern," whereas fetishism is equated with irrationality, superstition, and "primitive" religious beliefs.

To Marx's credit, it must be noted that he did not use fetishism—as did those anthropological writers from whom he borrowed the term—as a means of distinguishing "modern," scientific-technological (i.e., European or Western) thought from the "primitive" beliefs of "premodern" and non-Western cultures, but rather, as a device to point out the "primitivism" of Western culture and economics. At the same time, however, Marx's belief in the subject-object division and in the instrumental view of the world is so strong that, for him, any belief that imputes to objects—or to technology—an autonomous power, agency, or "life" could only be seen as a kind of mystification, as a return or regression to a more "primitive" form of belief. Marx was not, of course, alone in taking this view. Indeed, as Horkheimer and Adorno have suggested, the very notion of Western "modernity," of scientific "enlightenment," is founded on a demystification of the world in which those magical and animistic modes of thought (including what is pejoratively called "fetishism") that attribute a certain autonomy, agency, or life to the "objects" of the world are repressed in favor of a view that sees the world solely as the object of human knowledge and control.[2] This "disenchantment of the world," to use Max Weber's phrase, requires the figurative "death" of these objects—and, indeed, of the world itself. Otherwise, seeing the world and its "objects" as endowed with some form of "life" would bring the position of the modern (Western) subject, and its presumed mastery over the object-world, into question.

If this instrumental, rationalized view of the world defines "modernity," it is also intimately linked to the modern conception of technology. Indeed, as Herbert Marcuse has argued, the instrumental view of the world is essentially a "technological" view: "The science of nature develops under the *technological a priori* which projects nature as potential instrumentality, stuff of control and organization. And the apprehension of nature as (hypothetical) instrumentality *precedes* the development of all particular technical organization."[3] Marcuse's argument here is, of course, largely a restatement of Heidegger's interpretation of modern technology as an "Enframing" of the world in terms of its instrumental value. Thus, for both Marcuse and Heidegger, modernity is, by definition, "technological." Because, moreover, the notion of "the West" rests on the premise of its "modernity," on its "basis" in a "modern," scientific-technological rationality, technology also becomes the means by which "the West" distinguishes itself from the more "primitive" or, at best, "underdeveloped" (i.e., "not yet modern") beliefs and practices of the "Third World." Technology in this sense is not only figured as inherently "Western," but as fundamentally opposed to those "other" modes of thought that attribute a certain autonomy or life to the "objective" world. Indeed, the "modern" conception of technology is premised on the exclusion and repression of these kinds of "magical" or "fetishistic" beliefs, on the "death" of that "other" life that seems somehow "beyond" the technological Enframing of the world. "Modern technology," then, is supposed to be the very opposite of "fetishism"; as the essence of modern instrumentality, it—even more than the world that it "enframes"—is supposed to be "dead."

Yet, modernity and modern technology have always been haunted by an insistent return of the "dead," a return of what has been repressed by the instrumental or "technological" view of the world. It is precisely this return that is represented in the innumerable modern myths of technology come to life, where images of "Frankensteinian" technologies and ghosts in the machine serve to portray the threat to the human subject posed by an autonomous, uncontrollable technology. This technological coming to life is all the more frightening in that it involves a return to life of something that was supposed to be a "dead" object or instrument. It is therefore depicted not only as an inhuman, other life, but as one that is monstrous and uncanny.

As Freud defined it, the sense of the "uncanny" always involves a return of the repressed.[4] For Freud, moreover, this repressed is explicitly associated with "the old, animistic conception of the universe," which Freud sees as an earlier "stage of development" that has been "surmounted" by modern scientific-technological thought. The "uncanny" coming to life of machines and automata therefore represents a return (to "life") of that animistic or magical thinking repressed ("killed") by technological modernity—a return of a "technological other," of what might in fact be called the *technological unconscious.*

From a modern perspective, of course, this technological unconscious could only appear as an unsettling technological other, a monstrous or libidinal technology, technology represented as a chaotic, often destructive, force or movement. Frankenstein's creature, although he is not explicitly technological in the way that later robots, computers, and cyborgs will be, is in many ways the model for this coming to life of a monstrous technology. Unlike human life, the life of Frankenstein's creature cannot be figured as organic or whole. He is an uncanny assemblage of spare parts, the result of a mixed, "unnatural" reproduction, a kind of technological miscegenation. As such, he threatens the presumed mastery that has distinguished the modern "human subject" from its "others." Indeed, it is precisely Frankenstein's efforts to master the secrets of life that bring about this return of what modernity has repressed.

The Frankensteinian scenario—in which technology either comes to life and turns against its human creators or else provokes monstrous, mutant forms of life—will eventually become a standard formula for science fiction. At times, the libidinal aspects of this return of the technological repressed are made quite explicit, as in the case of the "Monster from the Id" in the 1956 film *Forbidden Planet.* This monster originates as a result of the highly advanced technology—built by an alien race, the Krel—that enables what Freud referred to as an "omnipotence of thought"; it allows its users to bring their thoughts into literal existence. Yet, when this technology is rediscovered by Dr. Morbius, it also projects into reality the destructive, libidinal forces of his unconscious, which take the form of an invisible, uncontrollable monster, bent on driving off or destroying all those who threaten his isolated relationship with his daughter. Here, then, the extension of technological rationality and control to its logical conclusion—thoughts technologically transformed into reality—also brings about a powerful return of what has been repressed by that rationality, represented here as "primal" urges. A similar pattern can be seen in Fritz Lang's *Metropolis* (which is more fully analyzed in chapter 2), where the repressive technological rationality of Fredersen, the "Master of Metropolis," leads to the unleashing of a chaotic and clearly libidinal technology in the form of the robotic False Maria.

There are, of course, many other examples of this Frankensteinian scenario in which modern technological rationality's attempts to control the world bring about a return of the repressed, of the technological unconscious.[5] Yet, *Forbidden Planet* and *Metropolis* are particularly useful examples in that they also make clear that such representations of a repressed technological other are associated not only with libidinal or sexual otherness—for example, the destructive "instincts" and vampish sexuality of the False Maria, Morbius's incestual desires and "primal" aggressivity—but often with cultural otherness too. Thus, for example, the sexual allure of the False Maria's dance is heavily dependent on its associations with "exotic" cultural elements. This dance takes place in a "club" known as Yoshiwara, which is referred to in the film as "that house of sin." The name itself therefore serves to suggest a connection between sexual or erotic pleasures and the "exoticism" of other cultures. When, moreover, the False Maria makes her appearance in Yoshiwara—wearing a vaguely Babylonian headdress and an exotic, diaphanous dress—she does so by rising from what appears to be a huge sacrificial dais, which is adorned with ornate gold reliefs and supported by Nubians clad in loincloths and wearing gold earrings. Her highly sexualized movements during this dance also seem designed to evoke a sense of some "decadent," ancient, and certainly foreign—perhaps Middle Eastern—culture.[6] Immediately following her dance, she reappears on the dais, now seated on the back of a huge, golden idol of a seven-headed hydra, the supporting Nubians transformed into stone figures of the seven deadly sins. The Hydra's legendary ability to regrow two heads for every one that is crushed or severed seems perfectly calculated, in this context, to represent the insistent return of repressed libidinal urges (or "sins"), while at the same time serving as a figure of technology's ability to double or reproduce itself—just as the False Maria doubles the "true" Maria.[7]

As a kind of technological fetish, which both replaces the "original" and hides the fact of its loss, the False Maria is here associated not only with sexual fetishism, but with "primitive" religious fetishism. This association of technology with a kind of religious fetishism is also present in Freder's vision of the "Moloch machine," where he sees the great subterranean machines of Metropolis—their underground placement itself serving as a representation of the technological unconscious—as an embodiment of the "primitive" rituals of human sacrifice and the "fetishistic" idol worship of the Tyrian god Moloch. In each of these cases, technological fetishism is represented as a kind of "primitive" monstrosity.

In *Forbidden Planet,* the connection of the technological unconscious to other cultures may seem less obvious at first glance. Yet, the film is loosely based on Shakespeare's *The Tempest,* with Morbius cast as a science-fictional Prospero, his daughter Alta as Miranda, Robby the Robot as Ariel, and the Monster from the Id taking the place of the deformed, half-human Caliban. Caliban has long been read as representing "man at his lowest, half-merged with the animal,"[8] but postcolonial scholars have pointed out the extent to which Caliban—whose name forms a nearly perfect anagram for "cannibal"—is also associated with non-Western cultures (particularly of the "New World"), and have in fact identified him with the continually defiant colonial subject repressed by European rule. Morbius in fact describes the Monster from the Id in terms that recall this heritage, calling it "the Beast, the Mindless Primitive" that lurks in the unconscious, repressed by the "modern," rational mind.

Like Caliban too, the Monster from the Id has mixed origins—its coming to life results from a combination of the human unconscious and alien technology. Indeed, it is described as a kind of impossible mixture of contradictory traits, contrary to the laws of nature. One character describes it as follows:

> This thing runs counter to every known law of adaptive evolution. Notice this structure here, characteristic of a four-footed animal, yet our visitor last night left the tracks of a biped. It's primarily a ground animal too, yet this claw could only belong to an arboreal creature, like some impossible tree sloth. It just doesn't fit into normal nature. Anywhere in the galaxy, this is a nightmare.

The Monster from the Id, then, becomes the figure of a "nightmarish" mutation and mixture of differences—differences so great that their combination can only seem "unnatural," "impossible," "counter to every known law," as though these elements could only be brought together by some violent force (much like Frankenstein's creature). Thus, the Monster from the Id serves as the representative of not only a "primitive," libidinous sexuality and violence, but of a kind of monstrous cultural and technological miscegenation—echoing Prospero's fears of Caliban's lust for Miranda, and the similar, racist fears of rape and miscegenation that Octave Mannoni has referred to as the "Prospero complex."[9]

As the False Maria and the Monster from the Id demonstrate, then, the technological unconscious is frequently cast in terms of a monstrous otherness—an otherness that is not simply technological, but also cultural and perhaps even racial. This linkage of technological and cultural others is in fact implicit in technological modernity's repression of those "other," "nontechnological" modes of thought that attribute a certain autonomy, life, or agency to the world. The uncanny technological life that animates both the False Maria and the Monster from the Id represents the return of this repressed "other" life, which modernity associates with "primal, libidinous urges" and "primitive, fetishistic beliefs," and which it inevitably figures in terms of a threatening monstrosity, mutation, miscegenation, and mixture.

The technological unconscious, then, is always a techno-cultural unconscious, in which technological otherness and cultural otherness are linked by virtue of the fact that both are excluded from and by Western technological modernity. Indeed, rather than viewing this modernity in Freudian terms of ego, repression, and the unconscious, one might well argue that Freud's division of the psyche is itself a replication of the "psychic" structure of technological modernity. Thus, modern technological rationality becomes the very model for the ego, while its repression of "primitive," "irrational," animistic, or magical beliefs figures the primary repression of "irrational," libidinal instincts through which the unconscious is formed.

In an age of high tech, however, these representations of the techno-cultural unconscious, particularly those of an autonomous technological life or agency, have begun to undergo a certain shift. Thus, for example, the monstrous computers, robots, and other technological mutants of traditional science fiction have given way to notions of artificial life, artificial intelligence, and intelligent agents, to cyborgs and other biotechnological life-forms that, despite their "inhuman" status, are no longer represented simply as a threat to humanity. Yet, neither are these new forms of technological life presented simply as unproblematic, utopian instruments or servants of humanity, as in the case of the robots in Isaac Asimov's stories or *Forbidden Planet*'s Robby the Robot.[10] Indeed, what defines this new sense of high-tech life is precisely the fact that, as in those "magical," "animistic," or "fetishistic" notions of life repressed by modernity, its life develops and mutates in ways that are often beyond rational human knowledge and control. Such technological life-forms may, therefore, still be regarded as "other," or even as "monstrous," but their monstrosity is, as Donna Haraway has suggested, replete with promises, with generative, mutative possibilities.[11]

The metaphors of mutation and mixture are in fact crucial to these new, "more promising" representations of high-technological life. Virtually every example of this kind of technological coming to life involves some type of mutation, some unplanned, unpredictable, or accidental combination of elements. Yet, this technological life is not only a result of mutation; it is *itself* figured as a kind of mutation: as a movement or process of change, of new combinations and mixtures, that remains beyond human prediction or mastery. As we have seen, there is nothing new in representing technological life in terms of an "unnatural" monstrosity, mutation, and mixture. What is new in these more recent representations of technological life is that mutation, mixture, and even perhaps monstrosity have come to be seen much more positively.

An illustration of how this shift in notions about technology, mutation, and monstrosity takes place can be found in what may seem an unlikely source: Japanese monster movies. The most widely known, and certainly the most enduringly popular, of these filmic monsters is Godzilla, who, beginning in 1954 with *Gojira* (an alternate version was released in the United States as *Godzilla: King of the Monsters* in 1956) and continuing to the present day, has starred in more than twenty films. Indeed, the success of the first Godzilla film spawned an entire genre of similar monster films, including not only the many Godzilla sequels, but also such films as *Rodan* (1957), *Mothra* (1962), *The Invincible Gamera* (1965), and their numerous sequels. Yet, if *Godzilla: King of the Monsters* established many of the conventions of this genre, it also derived many of its basic plot elements from the U.S. film *The Beast from 20,000 Fathoms* (1953), in which a giant dinosaur is thawed by atomic testing and proceeds to wreak havoc on New York City. Similarly, Godzilla (at least in the early films) is supposed to have been a type of dinosaur who was awakened and thawed by atomic testing—an origin story that would be repeated almost exactly for Rodan, the giant pterodactyl, Mothra, the giant moth, and Gamera, the giant prehistoric turtle.[12] As in *The Beast from 20,000 Fathoms*, then, these films serve as cautionary tales about the "disastrous" consequences of technological development, particularly atomic technology—an issue with obvious resonance in postwar Japan. This issue is further emphasized in the later Godzilla films, where Godzilla's existence is explained not simply as the result of his being *awakened* by atomic testing, but as a *mutation* brought about by his exposure to atomic radiation.

Obviously, then, Godzilla and the other monsters symbolize the "other side" of technology—the uncontrollable, destructive forces and mutations that modern technology can accidentally unleash. Yet, as the monster genre progresses, another curious accident or mutation occurs: the monsters turn from being threatening, purely destructive forces—as with the dinosaur in *The Beast from 20,000 Fathoms*—to being heroic figures, defending Tokyo and the Earth itself against other, more threatening monsters, and even against alien invasion. Through this shift, the idea of a monstrosity or mutation that is beyond instrumental human control comes to be seen not simply as threatening, but as potentially beneficial.

Some might argue that this shift is simply the result of the initial popularity of these monsters, which dictated their return (much as in *Terminator 2*) as "heroes" in the later films. This argument, however, begs the question of why they became popular in the first place, and why, for example, *The Beast from 20,000 Fathoms* did not spawn its own sequels. As in *The Beast from 20,000 Fathoms,* the Japanese monster films represent the return of a monstrous, premodern—indeed, "prehistoric"—life. Yet, in a way that *The Beast from 20,000 Fathoms* does not, they make explicit that this "return to life" is a return of precisely that "old" sense of a magical, animistic, or supernatural life that is repressed by modernity. Indeed, the Japanese word for these monsters, *kaiju,* carries connotations of the supernatural.[13] In the films themselves, moreover, one finds both Godzilla and Mothra worshiped as deities by their "primitive," native followers, whereas Gamera is revived from death by the prayers of children. Godzilla, too, seems virtually immortal, able to regenerate himself after his own death like some kind of atomic-age phoenix. Thus, the Japanese monsters, even as they represent the destructive forces and mutations that can be "brought to life" by modern technology, also symbolize the return of a "primitive," supernatural life force.[14] Yet, although this supernatural technological life proves beyond the control of modern science and technology, it is not represented simply as a dangerous, uncontrollable other.

It is suggestive that this representation of a supernaturally charged technological monstrosity first emerges in Japan, a country that is at once "technologically advanced" and "non-Western." Indeed, it could perhaps only have emerged in a culture that, to Western eyes, often seems to be a strange mixture of technological modernity and traditional (fetishistic?) belief in ghosts, spirits, and supernatural forces that are beyond human control. The intersection of these "modern" and "traditional" beliefs may account for not only the supernatural overtones attached to the notion of technological mutation in the monster films, but also the much more explicit connection of the technological and the supernatural in later Japanese *manga* and *anime.* In, for example, the *Super Dimension Fortress Macross* television series and the Americanized *Robotech* series, the secret of the advanced alien technologies seems charged with supernatural overtones. In *Akira* (1988), the force or life that results from technological mutation is very explicitly supernatural, beyond human control. Indeed, this mutant technological life, despite its monstrous and destructive aspects, seems to carry an almost messianic promise of rebirth and redemption.[15]

If, however, it is in Japanese sci-fi films that this more positive image of a supernatural, technologically mutated life first rears its monstrous head and slouches toward Tokyo, it does not remain exclusively Japanese for long. Indeed, it begins to spread, to replicate, to disseminate itself, as if this notion of technological mutation and life had acquired a monstrous, fetishistic life of its own. And, in a sense, it has. In other words, this emergence of a new, more positive vision of technological mutation and life also implies a "mutation" in the very notion of technology, which itself comes to be conceived less as a matter of "dead" instrumentality than as possessing a shifting, mutative, and perhaps supernatural life of its own.

This shift in the notion of technology can be seen in the many representations of technological mutation and life descended from the Japanese monster and sci-fi films and from Japanese *manga.* It is worth noting that many of these representations appeared, as with the Japanese films and *manga,* in popular cultural forms and genres that have generally been regarded as beneath the notice of those who give serious consideration to technology and techno-culture. Indeed, the narratives of technological mutation and life that occur in popular and mass culture are in many ways simply a thematization of the process of evolution and mutation that is already ongoing within mass culture. Given the close relation between mass culture and technological reproduction, it might well be argued that cultural forms are precisely the codes through which technological mutation takes place. In this sense, contemporary culture is, indeed, already science-fictional.

In fact, one of the earliest genres in which these representations of technological mutation would appear was science-fiction superhero comic books. This was especially the case with the mutant heroes of Marvel Comics, beginning in the early 1960s. Much of Marvel's success has been based on heroes whose powers were the unpredictable result of various technological accidents and mutations, among them the Fantastic Four, the Incredible Hulk, the Amazing Spiderman, and the most famous mutants of all, the Uncanny X-Men. In these comics—much as in Japanese monster films, *manga,* and *anime*—mutation, although it was still regarded as "uncanny," was for the most part presented positively; indeed, the sense of being a mutant (i.e., monstrous, different, other) often became a point of identification for the readers of these comics.

If, however, these comic books, like the Japanese monster films before them, presented the idea of technological mutation and monstrosity positively, this mutation was still a matter of living beings transformed by a technology—usually by some form of radiation—that was portrayed in negative, often dystopian, terms. Yet, once these positive portrayals of technologically altered mutants had appeared, it was perhaps only a matter of time before the idea of a mutant technological life—of technologies that had themselves mutated and come to have a life of their own—was also represented in more positive terms. Indeed, during the 1980s, these new representations of a mutant technology come to life suddenly seemed to appear everywhere at once: in popular science-fiction literature and films such as *Videodrome* (1983), *Neuromancer* (1985), and *Max Headroom* (1985), in theoretical discourses such as Donna Haraway's "A Manifesto for Cyborgs," Gilles Deleuze and Félix Guattari's *A Thousand Plateaus,* and Avital Ronell's *The Telephone Book,*[16] in technological discourses on self-replicating computer programs (from discussions of computer viruses and worms to debates over artificial intelligence and artificial life), and in scientific discourses on nonlinear and "chaotic" dynamics. In all of these cases, the "emergence" of technological life has been characterized as a kind of mutation: the result of unpredictable, and often accidental, combinations or mixtures of elements.

At times, these mutations—through which technology is brought to life—are portrayed quite literally as accidents, as in the case of the personal computer in the film *Electric Dreams* (1984) that attains sentience when its circuitry is altered by a spilled soft drink or the robot in *Short Circuit* (1986) that comes to life when struck by lightning. Such accidents are, however, presented as the singular result of forces or events that are external to the technology in question. These technological life-forms may have achieved the status of independent, living beings, but they did not do it on their own, nor is such a mutation likely to recur (except in sequels). Increasingly, however, one sees the mutation from technological object to technological life represented as a process of "growth" or "evolution" that is internal to technology, that is *itself* technological.

This "evolutionary" view of technological life is well illustrated in the Internet slogan "The Net wasn't built; it grew." In this slogan, the modernist privileging of "built," technologically inspired designs over more organic models is reversed.[17] Here, in fact, technology itself becomes "organic," living: no longer dependent on human planning and construction, it grows in unpredictable ways, evolving its own structures according to what seems to be its own internal logic, its own secret "aesthetic."

This "aesthetic" is an aesthetic of pastiche—which is to say, an aesthetic of complexity. Indeed, the idea of complexity, as it has been elaborated in recent scientific discourses (including what has popularly become known as "chaos theory"), is central to contemporary notions of the emergence and evolution of technological life. This is particularly evident in the newly developing field of computer-generated "artificial life," where, as Sherry Turkle has noted, one of the "defining features" of "the emergent aesthetic of artificial life" is that "complex behavior can emerge from a small number of simple rules."[18] Turkle's point is in fact echoed in the instruction manual for the computer game SimLife, which is intended to provide "an exploration of the emerging computer field of Artificial Life": "One of the most important features of A-Life is emergent behavior—when complex behavior emerges from the combination of many individuals following simple rules.... Another important aspect of A-Life is *evolution*—artificial life forms can react to their environment and grow, reproduce and evolve into more complex forms."[19] Thus, it seems, the emergence and evolution of complexity—in terms both of behavior and of form—becomes the defining characteristic of artificial or technological life. What is it, however, that defines this "complexity"?

Steven Levy, in his book on artificial life, explains that a complex system is "one whose component parts interact with sufficient intricacy that they cannot be predicted by standard linear equations; so many variables are at work in the system that its overall behavior can only be understood as an emergent consequence of the myriad behaviors embedded within."[20] Complexity is therefore a matter of unpredictability: when bifurcation and combination within a system reach a certain level of complexity, the results that emerge from these interactions become impossible to predict or control.[21] Such a system must therefore be seen, like life, as autonomous, as governed by its own internal processes of replication and mutation, from which it evolves its own patterns, organizations, and behaviors. It is precisely this ability to mutate and evolve autonomously—the ability to "self-organize"—that allows complex technological systems such as the Internet to be seen as living.

It is important to note that the aesthetic of complexity that underlies this sense of an emergent technological life remained largely unnoticed—at least, in Western, scientific-technological discourse—until the advent of high technology. It was, in fact, only with the proliferation of digital computers and multiprocessing that complex, nonlinear systems could be given a visual—that is to say, an aesthetic—representation (as in, for example, the case of fractals).[22] Yet, if high tech enables the visualization of this aesthetic of complexity, it is also the case that high tech is in many ways defined by an aesthetic of complexity. What distinguishes high tech from the modern notion of a predictable and controllable instrumental technology is precisely its complexity. The aesthetic and fetishistic qualities associated with high tech are also a result of this aesthetic of complexity, of the sense that technology has achieved a complexity that is beyond the ability of humanity to predict or control. It is, therefore, through the emergence of this aesthetic of complexity that high tech has increasingly come to be seen as having—like the artistic or fetish "object"—an autonomy, a life, of its own.

This idea of a complex, uncontrollable, and autonomous technological life is in many ways fundamentally opposed to Western modernity's instrumental view of technology and of the world. The sense of high tech's complexity may in fact explain the frequent association of high tech with the idea of "postmodernity," which has also, quite often, been defined in terms of mixture, pastiche, and complexity. Yet, it may also explain why there has often seemed to be a special affinity between high-tech discourses and those magico-spiritual modes of thought regarded by modernity as "premodern," "superstitious," or "primitive." The idea of an autonomous technological life emerging from the unplanned combinations and interactions of complex technological systems suggests the existence within technology of a generative, mutational force or movement that is beyond human mastery. Here, technology is clearly no longer being conceived in terms of an instrumental rationality, but as something that has its own complex, inscrutable logic, its own "mystery." Indeed, as Turkle has observed, these complex technological phenomena "can seem magical. . . . They resonate with our most profound sense that life is not predictable. They provoke spiritual, even religious speculations" (pp. 166–67).

This sense of a magical or spiritual technology involves, without question, a kind of fetishism, but it seems to be less a matter of commodity fetishism than of a return of a magical or supernatural thinking. Contemporary high-tech discourses are, in fact, often permeated by magical or spiritual terms and ideas. In a number of cases, these terms and ideas have been drawn from the science-fictional and fantasy worlds of novels and role-playing games such as Dungeons and Dragons and Magic, worlds in which magic and sorcery play a prominent role. Thus, for example, terms such as *avatars, wizards, demons,* and *sprites* have become a standard part of the computer networking and "mudding" lexicon. Indeed, the terms *MUDs* and *mudding* were themselves derived from this same source: MUD—a term used to describe a virtual computer space where multiple users can enter and interact—was originally an acronym for Multi-User Dungeon.

In his science-fiction novella "True Names," Vernor Vinge extends this magical metaphor inherent in MUDs so that its applicability to computer networks and programming becomes clear: in "True Names," all of the world's networked data becomes a "magical realm" where seeing computer programmers as wizards and their programs as magic spells is "more natural than the atomistic twentieth-century notions of data structures, programs, files and communication protocols."[23] In a similar vein, Malkah, a character in Marge Piercy's novel *He, She and It,* observes the similarities between Jewish mystical thought and "the world of artificial intelligence and vast bases in which I work—the world in which the word is real, the word is power, energy is mental and physical at once and everything that appears as matter in space is actually immaterial. Perhaps that's why as I get older, I become more of a mystic."[24]

In both Vinge's and Piercy's narratives, magic—and thus technology—tends to be presented as a means of empowerment and control, much as it is in Dungeons and Dragons and other fantasy narratives and games. Certainly, much of the allure of computers and computer networks has been based on precisely this promise of empowerment, of mastery, whether it has been presented as "magical" or not. Scott Bukatman has noted this tendency toward the empowerment of select individuals in "True Names" and has argued that it is "continued" in William Gibson's *Neuromancer,* particularly in the sense of freedom and power that Gibson's cyberspace cowboys gain from their mastery of the electronic realm.[25] Bukatman, in fact, explicitly links this sense of cyberspatial mastery to Freud's notion of the uncanny and to "the return of the animistic view of the universe *within* the scientific paradigm" (p. 210). Thus, Bukatman argues that "the penetration of consciousness into the cyberspatial matrix is an extension of the power of the will which recalls the 'animistic' conception of the universe that precedes the emergence of the mature ego" (ibid.). Bukatman is not alone in citing Freud here; Mark Dery makes a similar argument about "modern primitivism," which he describes as "the recrudescence, in computer culture, of the 'primitive' worldview—'the old, animistic conception of the universe,' with its 'narcissistic overestimation of subjective mental processes.'"[26]

Thus, even as Bukatman and Dery point out the return of a magical or animistic thinking in contemporary techno-culture, they follow Freud in assuming that this sort of worldview necessarily involves an "omnipotence of thoughts" and a refusal of limits on the individual subject. In so doing, they accept—presumably unconsciously—Freud's "developmental" and highly Eurocentric version of history, in which "the scientific view of the world" is equated with a "maturity" that "no longer affords any room for human omnipotence."[27] There is considerable evidence, however, that magic and animism are not simply a matter of asserting one's thoughts or will upon the world, but are instead a way of interacting with forces that are seen as *beyond* human control. Indeed, one might well argue that it is "scientific" thinking that sees human beings as potential gods, able to know and control everything around them. Moreover, although it is certainly clear that the empowerment of the individual subject is crucial to many contemporary discourses on technology, we should not assume that this fantasy of self-empowerment is the only fantasy operative in these cases. Bukatman, for example, supports his argument by citing Claudia Springer's assertion that "The pleasure of the interface, in Lacanian terms, results from the computer's offer to lead us into a microelectronic Imaginary where our bodies are obliterated and our consciousnesses are integrated into the matrix."[28] Yet, Springer actually argues that fantasies of incorporation in cyberspace, or in technoculture more generally, are not simply a matter of "self-affirmation," but also involve the "seductive" possibility of losing oneself in something beyond one's control (pp. 59–62).

Bukatman, to his credit, does acknowledge that, in the course of Gibson's trilogy, the role of individual cyberspace jockeys diminishes as the representation of the matrix itself changes (p. 214). In the *Neuromancer* trilogy, the "unthinkable complexity" of the matrix leads to a mutation, referred to as "When It Changed," from which emerge autonomous, sentient artificial entities that no individual can know or control. Beyond human understanding, these "ghosts in the matrix" take on a spiritual or religious dimension; indeed, Gibson represents them, quite literally, as gods.

Here, we might seem to be dealing with not simply a fetishism of technology, but an attempt to respiritualize the technological, much like the attempts of artistic modernism to reinvest the technological world with a sense of aura, of eternal values. Yet, unlike those modernist attempts to mediate and bring together the modern and the eternal, the scientific-technological and the aesthetic-spiritual, Gibson's gods in the matrix do not seem designed to provide a sense of cosmic unity or eternal values. If there is an aura to these gods, it is an aura of otherness rather than unity, of process rather than presence.

Gibson chooses, in fact, to represent this "technological spirituality" in "other," non-Western terms; his gods in the matrix manifest themselves in the forms of various loa common in Afro-Haitian voodoo. Thus, Gibson explicitly links cyberspace, artificial intelligence, and, by extension, digital technology generally, to a "return" of precisely those animistic or magical beliefs repressed by modern scientific-technological thinking. Gibson has in fact argued that "The African religious impulse lends itself to a computer world much more than anything in the West. You cut deals with your favorite deity—it's like those religions already are dealing with artificial intelligences."[29] Along similar lines, John Perry Barlow, a prominent Internet activist and theorist, has suggested that Africans, as well as "others in the so-called developing world," may have a cultural advantage in coming to terms with "digital society," precisely because they have not been caught up in modern, industrial thought.[30] Although Barlow does not explicitly mention "spiritual" or "religious" beliefs in his analysis, his argument rests on an implicit linkage of pre-industrial and postindustrial thinking. In this, he echoes the techno-tribal rhetoric of Marshall McLuhan, who argued that "the new electric technology is retrogressing Western man back from the open plateaus of literate values and into the heart of tribal darkness, into what Joseph Conrad termed 'the Africa within.'"[31]

These figurations of computer and electronic technology in terms of a pre-modern, non-Western, or "other" way of thinking suggests a shift in the conception of technology itself, and in humanity's relationship to this technology. As technology becomes more complex, more dense, and less comprehensible in its entirety, it begins to "appear" as an autonomous, uncontrollable, even supernatural other. As such, technology cannot be conceived simply as an instrument or object under human control. Rather, humanity's relation to technology becomes much more akin to the relation to spirits or gods in "magical" or "animistic" belief systems, which Gibson describes as a matter of interaction, cooperation, of making "deals."

These kinds of "magical" or "spiritual" figurations of technology can in fact be seen in a host of contemporary technological movements and discourses, including "modern primitivism," "new-edge" science, techno-shamanism, and techno-paganism, which frequently draw on magical, spiritual, and metaphysical discourses to figure a new relation to technology, to a techno-cultural world that often seems out of control. In an article on techno-paganism in *Wired*, for example, Erik Davis argued that the representations of computer technology favored by techno-pagans are based on magical and other pre-Enlightenment modes of thinking that allow the "spiritual potential" of technology to be "taken seriously."[32] Davis, much like Gibson, also suggests that this tendency to see technology in magical, spiritual, or supernatural terms is increasingly appropriate to the complexity of the contemporary technological world: "As computers blanket the world like digital kudzu, we surround ourselves with an animated webwork of complex, powerful, and unseen forces that even the 'experts' can't totally comprehend. Our technological environment may soon appear to be as strangely sentient as the caves, lakes, and forests in which the first magicians glimpsed the gods" (p. 177).

Here, as in Gibson's work, not only does technology come to be seen as an autonomous, uncontrollable, sentient other, but this technological other is figured in precisely those terms that Western modernity has defined itself against, for modernity has always regarded the magical, the spiritual, and the supernatural as its other, as irrational, premodern, non-Western, and so on. Technology, which had been considered the very basis of human knowledge of and control over the world, here becomes a figure of the limits of human mastery, of that which is "beyond" human knowledge and control. It becomes, to use more traditional philosophical terms, a figure of the sublime, a term whose religious and aesthetic connotations seem entirely appropriate to the growing sense of the incomprehensible size and complexity of the contemporary techno-cultural world.

From a modern perspective, however, the idea of a technology that is uncontrollable, sublime, or godlike not only challenges the status and mastery of the human subject, it evokes the fear that human beings will themselves become mere objects or tools: slaves to technology, cogs in the machine, technologized zombies. From this point of view, it seems, to be less than "a master" is to lose one's status as a human subject—to become, in other words, less than human. Yet, although the rhetoric of many contemporary technological discourses continues to figure technology solely in terms of its ability to empower human beings, to perfect their mastery of the world, there is also, as the examples of Gibson and the techno-pagans demonstrate, a growing tendency to view the relationship between human beings and technology in terms other than those of master and slave, subject and object. Although the technological "gods" that appear in these discourses may be "beyond" human control, this does not mean that there can be no interaction with them. It does not mean that human beings become the slaves of technology, that they lose all sense of autonomy and agency. It merely means that humans must deal with the technological world not as a series of tools or objects, but as a host of autonomous forces or agencies.

It is hardly surprising that the attempts to depict this new sense of technology would draw on precisely those magical, spiritual, and supernatural discourses repressed by modernity's instrumental view of the world, for these discourses provide models for dealing with the world from a position other than that of control and mastery. There are, in fact, many examples in contemporary discourses on technology where the relation of humans to technology is described in terms that suggest communion with supernatural forces, hypnotic or trance states, or other forms of openness to the ineffable complexities of a technological world. Technology, in these cases, comes to be seen as an ongoing movement or force that human beings can, without necessarily understanding it, take part in, even become part of. Often, in fact, this notion of technology is figured in fluid or oceanic terms, as something that people can not only immerse themselves within, but set themselves adrift in, abandoning themselves to the seemingly random complexity of its currents or waves. Thus, for example, one can understand the popularity of the various "wave" and "surfing" metaphors that have so often been used to describe people's relations to technology. Similarly, this sense of self-abandon, of allowing oneself to "let go" of mastery and become part of a larger force or movement, and to be carried along by it, explains a good deal about the "techno-shamanistic" aspects of such discourses as rave, techno, and "trance" music and culture, where participants use electronic music, computer graphics, and, often, synthetic hallucinogens to "lose themselves" in trancelike states. Similar impulses also seem to underlie what *Mondo 2000* has called "techno-primitivism," where the appeal of computers and electronic technologies is combined with the attraction of such subjects as body piercing, tattooing, drumming circles, and exotic hallucinogens, yielding discussions of such curious amalgams as holographic tattoos, electronic dental implants, and computer programs designed to induce trance states. Yet, however exoticizing or even silly some of these examples may seem, their use of elements drawn from discourses repressed by Western modernity is based on a recognition that, in a high-tech world, humanity's relation to technology is, or should be, less a matter of control than of openness and participation, of "letting go" of the sense of mastery that has defined the modern human subject.

This new relationship to technology, to the technological world, is not, then, a relationship between subjects and objects, but a relationship with others, among others. It is therefore no coincidence that the attempts to describe this relationship are often based on "other" discourses, discourses that are often associated with the "Third World" or with women.[33] Thus, for example, Gibson draws on Afro-Caribbean religion, and techno-paganism relies on a spiritual discourse that tends to emphasize the powers of women.[34] In these discourses, human beings are not defined solely in terms of their control of an objectified, instrumentalized world; rather, they tend to be seen as participants in a world filled with autonomous forces and agencies, with which they must deal, interact, cooperate. Nor, in these discourses, are the boundaries between the "human subject" and the "object-world" considered rigid or impassable. Instead, there is often a sense of openness, of connection and interconnection, of mixture, in the relations of humans and the forces of the world. In many ways, then, the "human" becomes a kind of hybrid, a permeable, mixed, and complex entity, a "monster."

In a high-tech world, of course, such hybrid monsters are called cyborgs. As theorized by Donna Haraway, cyborgs are not merely a mixture of the "human" and the "technological"; rather, the cyborg represents a "different" positionality for humans, an other position that stands in contrast to the Western, patriarchal notion of the human subject or "self," with its fixed boundaries and sense of mastery. Haraway, in fact, explicitly links the position of the cyborg to the situation of postcolonial women and women of color and to the positioning of what Trinh T. Minh-ha has called the "inappropriate/d other," which, as Haraway notes, "refer[s] to the historical positioning of those who cannot adopt the mask of either 'self' or 'other' offered by previously dominant, modern Western narratives of identity and politics."[35] What Haraway calls "the cyborg subject position" is, then, defined by this ambiguous partiality and hybridity, by the permeability of its boundaries, by its ability to exert a sense of productive agency that is not based on autonomy and mastery but on a relationality that challenges the dualisms of subject and object, self and other, male and female that have defined Western modernity.

Cyborgs, then, are "networked" entities; they do not exist simply as autonomous individual subjects, but through connections and affinities, including their connections to technology. Indeed, cyborgs are never entirely separate from technology, from the complex techno-cultural world in which they live. Neither the masters of technology nor the victims of it, they are participants in the chaotic mix of techno-culture, surfing its unpredictable waves and flows, taking part in its recombinant processes of (technological) reproduction, mixing random elements, unsettling and recontextualizing old meanings, generating new codes and new patterns of complexity from within technology itself. To be a cyborg, then, is to take part in a process that is very similar to what Heidegger called "bringing forth"; it is to participate in a technological *poiesis* that remains open, unsettled, and unsettling, irreducible to instrumental terms, whose results can never be known or controlled in advance.[36] Thus, in—or through—the cyborg, technology once again comes to be seen, as Heidegger hoped, as an ongoing, generative process or agency that opens rather than enframes, a process that is closer to aesthetic production than to the instrumentality of modern technology. For cyborgs, in other words, technology becomes technē—high technē.

An exemplary figuration of this idea of technology as a generative, "aesthetic" process can be found in Gibson's *Count Zero,* where he describes an artificial intelligence that devotes itself to the creation of shadow-box collages, assembled of bits and fragments drawn from a vast, chaotic mixture of discarded objects swirling in the zero gravity of an abandoned space station. In Gibson's description, this stockpile of objects is filled with "uncounted things," seemingly arrayed at random: "A yellowing kid glove, the faceted crystal stopper from some vial of vanished perfume, an armless doll with a face of French porcelain, a fat, gold-fitted black fountain pen, rectangular segments of perf board, the crumpled red and green snake of a silk cravat . . . Endless, the slow swarm, the spinning things . . ."[37] Yet, even as Gibson seems to emphasize the aleatory quality of this process, he also suggests that an unknown, and perhaps unknowable, logic is at work. Thus, his character Marly finds herself "hypnotized" by the movements of the AI's mechanical limbs: "as they picked through the swirl of things, they also caused it, grasping and rejecting, the rejected objects whirling away, striking others, drifting into new alignments. The process stirred them gently, slowly, perpetually" (p. 225).

The technological process that Gibson describes here is clearly recognizable as a form of artistic production: a "bringing forth" that unsettles and reassembles the elements of the world in new and unforeseen patterns. Here, however, art is no longer "aesthetic" in the traditional sense of romantic aesthetics. If, in this case, there is still a "hypnotic" sense of an aura, of something living, something that can "look back at us," this is an aura that is not based on a sense of wholeness, on eternal values. Rather, it is precisely a matter of partiality, contingency, hybridity, mixture. In this sense, the aesthetic, like the technological, has become a matter of style. As such, it becomes indistinguishable from culture more generally—or rather, from the unsettling processes of techno-culture.

Indeed, Gibson's description of this techno-artistic process can be seen as a description of the processes of techno-culture itself. The techno-cultural world is, in fact, an immense and complex reservoir of cultural images, objects, and stories, which are constantly being "stirred," unsettled, reproduced, mixed, altered, and recombined in ways that are simply too complex to be predicted or controlled. This process might be seen, in Heideggerian terms, as the "essence" of techno-culture, which has always exceeded the various attempts to "regulate" it, to make it useful, profitable, instrumental. Gibson's characters, however, tend to exhibit a certain openness to this technological process or movement; indeed, many of them seek to take part in it, or in fact become part of it. These characters can be considered cyborgs, not just in the sense that they have technological implants and prosthetics in their bodies or brains, but in the sense that they find themselves "incorporated" in the *processes* of technology, in the complex and only dimly understandable movements and forces that constitute the techno-cultural body.

The cyborg's incorporation in the processes of the technological body involves giving up some of the traditional sense of what it means to be "human," giving up—at least in part—the sense of individual autonomy and mastery that has defined the human subject throughout modernity. The cyborg body is, by definition, "inhuman," alien, monstrous, other; it has its own autonomous logic and movements; it is never entirely "one's self." Yet, to become part of this technological body does not mean losing all sense of individual agency; it does not mean becoming simply part of a machine. The situation, as Haraway might say, is "messier than that":[38] the interactions within this body are more fluid, more contingent, more complex than the opposition between "human subject" and "technological object" will allow.

It is precisely these messy complexities that are explored, as Haraway herself has noted, in the science fiction of Octavia Butler, particularly in the novels of her "Xenogenesis" series, where humanity undergoes a genetic merger with an alien species, the Oankali.[39] The Oankali are not technological in a conventional sense; they are, instead, the ultimate genetic or "biotech" engineers. They exist through genetic manipulation, mutation, merger, and exchange. Indeed, the drive to exchange and alter genetic material is part of their identity as Oankali. As one of the Oankali puts it, "We *must* do it. It renews us, enables us to survive as an evolving species instead of specializing ourselves into extinction or stagnation" (p. 39). Using these genetic engineering abilities, the Oankali live within a complex biotechnological environment; the entire Oankali culture, including even the spaceships in which they live, is based on biological, living technologies. The Oankali, as Haraway observes, "do not build nonliving technologies to mediate their self-formations and reformations. Rather, they are complexly webbed into a universe of living machines, all of which are partners in their apparatus of bodily production."[40] The Oankali, moreover, want to "incorporate" humans in this biotechnological web, to engage in a genetic exchange that will alter humanity and produce new, hybrid forms of life.

Butler's presentation of this "xenogenesis" contrasts sharply with the xenophobia that has generally characterized science-fiction portrayals of alien "incorporations." As an example of this contrast, one need only compare Butler's depiction of the Oankali with *Star Trek: The Next Generation*'s portrayal of the "Borg," a hypertechnologized race of alien cyborgs who threaten to "assimilate" the human race.[41] Although the Borg are cyborgs—a combination of living beings and technology—they are presented as mechanical, robotic. This mechanical status is evident, first of all, in their bodies: in the metal tubing that protrudes from their flesh and in the prosthetic devices that substitute for various limbs and organs. More important, though, the mechanical status of the Borg is demonstrated by their lack of individual emotions, agency, or will. To be assimilated by the Borg, as in fact happens to Captain Jean-Luc Picard, is to cease being human, to become a mere object or tool, a cog in the Borg machine, a slave. Assimilation by the Borg, then, is presented as a horrifying loss of control, a loss of the human subject's (i.e., the white European male's) position of mastery, neatly personified in the figure of Captain Picard. Indeed, much of the horror of this incorporation derives from the fact that it is presented as a violation of the boundaries of the autonomous masculine subject. It is suggested, in fact, that Captain Picard's assimilation by the Borg is the psychological equivalent of rape.

Butler, in many ways, highlights these same issues of autonomy—of slavery and violation of individual boundaries—in her narrative. Yet, her treatment of them, and of the relationship between humans and the Oankali, is considerably more complex. Although the Oankali are not technological in the sense that the Borg are, and therefore do not threaten the same roboticized loss of agency, humans still find their otherness—particularly their Medusa-like sensory tentacles—profoundly disturbing.[42] They see the Oankali, quite literally, as monsters. Yet, what these humans see as even more disturbing is the threat that genetic merger with the Oankali poses to humanity, and that mating with individual Oankali poses to the individual human subject. Mating with the Oankali is seen as participating in a monstrous, "unnatural" mixing, as taking part in a kind of miscegenation. Thus, humans who mate with the Oankali are perceived—and often perceive themselves—as traitors to humanity, and are frequently referred to as "whores." Yet, the Oankali, despite their otherness, are difficult to resist: contact with the Oankali third sex, the *ooloi,* through whom all reproduction takes place, stimulates the pleasure centers of the human brain and produces a biochemical bonding. In this way, humans and Oankali become connected, interdependent, at a physical as well as a psychological level. Butler's human characters find it difficult to understand how their conscious will, so often taken to define the autonomous human subject, is overcome by these unconscious, biochemical desires. Often, they feel that their personal boundaries have been violated, that they have lost their sense of autonomy. In Butler's portrayal, the characters who have the most difficulty with this situation, who commit suicide or become mad, are precisely those who are used to being in a dominant position—white males, in particular. Women and people of color, Butler suggests, are better able to cope with the Oankali "incorporation" precisely because their sense of identity is not so heavily based on maintaining rigid boundaries between subject and object, on a position of dominance or mastery. Already hybrid, already cyborgs, they are better equipped to interact with, deal with, and live with other forces and agencies, with other processes, that are to some degree beyond their control.

In Butler's narratives, these other forces and processes are not only alien, they are also technological, or biotechnological. Thus, Butler's presentation has considerable resonance in the techno-cultural world of high tech, where bits of cultural data often seem to take on a technological life of their own, propagating themselves through reproduction and dissemination, and undergoing mutation in the process. Noting the similarity of the process of cultural "replication" to genetic replication, the biologist Richard Dawkins has referred to these bits of cultural data, analogous to genes, as *memes.* The meme, as Dawkins notes, is not only a "unit of cultural transmission," but "a unit of imitation," which is "related to 'memory.'"[43] As such, the meme would seem to be, by definition, a matter of copies, of simulation. Thus, the rise of technological reproducibility seems to enhance the ability of memes to replicate and spread, to mutate and evolve. Not surprisingly, then, the notion of memes has itself been widely disseminated on computer networks, where the concept of "memetics" has seemed particularly suited to the rapid replication and evolution of digital information, which often seems to have a mutant technological life of its own.

Both the genetic technologies of Butler's Oankali and the techno-cultural life of memes suggest a conception of technology that is less a matter of hardware than of software, less a matter of mechanics than of information "processing." Technology, in other words, becomes an autonomous process through which information—including cultural and genetic information—reproduces, mutates, and evolves. Thus, as the world has become increasingly involved in this technological process, increasingly digitized, it has also come increasingly to be seen as a matter of data. Yet, the world or space of this data is not physical but virtual; indeed, as both computer discourse and Dawkins's concept of memetics suggest, it is a space of memory, of what might be called *techno-cultural memory*. We might imagine this techno-cultural memory as an infinitely more complex version of the current World Wide Web: as a networked memory from which all of the data of the world is accessible. Indeed, William Gibson has already imagined such a network in his depiction of a cyberspace "matrix." Yet, in many ways, we have already entered the "matrix" of techno-cultural memory, even if it is not yet hardwired. It is simply the sum of all the data that surrounds us in a techno-cultural world. Indeed, the space of this memory is techno-culture itself.[44]

Yet, as the techno-cultural web around us grows ever larger, it becomes increasingly complex; the interactions that take place among its elements become increasingly difficult to predict or control. At a certain point, then, it seems to take on a life or agency of its own, to evolve its own dynamics, its own logic, its own processes. Increasingly, as exemplified in the work of Butler, Haraway, and Gibson, as well as in a host of divergent discourses, from techno-paganism to A-Life to memetics, we see attempts to represent this sense of technology, of techno-culture, as a complex, self-replicating process that is continually mutating and evolving, continually generating new and unpredictable patterns, combinations, mixtures. In these representations, technology becomes something that unsettles or disorganizes the conventional boundaries between subject and object, self and other, the human and the technological. In fact, these representations tend to present human beings as inextricably involved or incorporated in the complex processes of technology that surround and infuse them. In, for example, Haraway's notion of cyborgs and Butler's presentation of a hybrid xenogenesis, technology is neither an external instrument nor simply a threatening, uncontrollable other, but a promising, generative process that, however monstrous or alien it may seem, is already ongoing within us. Thus, although Butler represents this technological process in biochemical or genetic terms, it bears a striking resemblance to the processes of the Freudian unconscious; for Freud's unconscious is itself a complex, unsettling process that, although repressed, goes on largely autonomous of the conscious human will. It too seems to have a life or agency that is at once uncontrollable, alien, other, yet dynamic, mutational, generative. It has its own logic (the logic of the signifier), its own (primary) processes, its own complex, mutational aesthetic, which cannot, by definition, be entirely understood or controlled. It has, in fact, often been cast in terms of an unpredictable fluidity and mixture, as well as in terms of complex webs and dense connections. And, of course, it works with, or through, fragments of memory, replicating and mixing elements, altering bits of data, to produce new images, patterns, and figures, which often seem to be charged with an uncanny, monstrous, supernatural life.

In a high-tech world, the complex processes of technology, of the techno-cultural world, have in fact come to seem more and more like the processes of the unconscious. Yet, this similarity may be less a matter of technology's resemblance to the Freudian unconscious than of the fact that Freud's unconscious was always technological in this sense. Thus, as suggested earlier, it is modernity's "repression" of the idea of technology as an autonomous process that becomes the model for the unconscious. Modernity's attempts to regulate and control these technological processes, to technologize the world in the interests of the human subject's autonomy and mastery, have only led to the increasing complexity of the techno-cultural world, and the increasing difficulty of controlling it. What is repressed, it seems, does return.

For some, this idea of a technological or techno-cultural unconscious will seem like yet another form of "fetishism," a semireligious mystification that hides the economic and social inequities of multinational capitalism. From this point of view, the processes of the techno-cultural unconscious will seem simply another name for the processes of capitalism itself; the idea of an autonomous technology will seem to be merely another version of Adam Smith's "Invisible Hand." And certainly, it would be pointless to deny that techno-culture and multinational capitalism are deeply imbricated in one another or that many corporations—particularly high-tech corporations—do tend to fetishize high tech and even the movements of techno-culture itself. Yet, to reduce the processes of techno-culture to a matter of capitalist instrumentality, to a matter of accumulation strategies and stratification patterns, is to do exactly what capitalism itself does. It is to try to resecure or reinstrumentalize, to control, the unsettling processes of techno-culture. Indeed, to confuse techno-culture with capitalism in this way is itself a species of fetishism, in the sense that it substitutes a fixed notion of instrumentality, accumulation, and control for more complex processes. As Haraway has argued concerning the techno-scientific fetishism of genes and gene mapping, these kinds of fetishes "make things seem clear and under control."[45]

Certainly, contemporary high-tech discourses often participate in this kind of techno-capitalist fetishism. For many high-tech corporations, as well as for such magazines as *Wired* and *Mondo 2000,* the fetishism of techno-capitalism becomes a means of entrenching their own positions of control and mastery. They seem, in fact, to view themselves as a kind of technological and "entrepreneurial" elite, a priesthood that worships at the shrine of technological advance and, not coincidentally, capitalist advancement. They see themselves, in short, as a *high-tech avant-garde*: brave souls who dare to surf the wave of high tech, to move fast enough to keep themselves on the "cutting edge" of technological innovation. In these terms, being on the "leading wave" becomes a way of reconfirming one's mastery of techno-culture, one's elite or dominant position in the world, and often, a way of reaping a profit from that position. It is, in short, a means of resecuring an autonomous techno-culture, of retaining a sense of human (or capitalist) autonomy and control over the complexities of the techno-cultural world. In this sense, the attitudes of this kind of high-tech vanguardism are often strikingly similar to the Italian Futurists' self-serving glorification of autonomous technology, a glorification that was often both profoundly aggressive and deeply misogynistic. Indeed, one need only glance at tracts such as *Wired*'s *Mind Grenades: Manifestos from the Future* or Bill Gates's *The Road Ahead* to confirm the high-tech avant-garde's tendency to view technology from the standpoint of a few, "select" masculine subjects.[46] To be in this technological vanguard is to stand in a privileged relation to technology, to be closer to its "pulse." And it is precisely this privileged status, this "higher access" to technology, that supposedly affords the high-tech avant-garde a privileged view of the future, as reflected in the titles of these books.

Here, high tech is itself figured as a matter of "height," as a privileged position that allows one to see further and know more. By having access to these "heights," the high-tech avant-gardists become the prophets, the seers, of the future—a future in which, they assure us, technology will empower everyone and thus promote social unity. It is perhaps worth recalling here the ways in which National Socialism also mobilized figures of "height"—for example, the myths of mountaintops and eagles—to promote its vision of the future, of an empowered and unified Germany. Indeed, *Triumph of the Will* begins with shots of Hitler's airplane amid the clouds and ends with Nazi soldiers marching upwards into the sky, into the future. In a sense, the Nazis attempted to appropriate the aesthetic figure of the sublime—and its association with heights and mountains, with the unknown and uncontrollable—to their own view of the future, giving it a direction, a road ahead, an end. And although, without question, the high-tech avant-garde should not simply be equated with National Socialism, its representatives do tend to appropriate the figure of technological height, of a high-technological sublime, as support for their own status as visionaries. They become the privileged interpreters of technology and the future, the priests of the high-tech "cathedral of the future." They become, in short, the mediators of technology, the "interface" through which the unknown, "sublime" complexity of high tech—and, indeed, the future itself—can be accessed by the general populace.

But, then, the myth of the future—of having special access to the future—has often been the basis on which various avant-gardes, and other "leaders," have attempted to privilege their own "height," their own power. One of the paradoxes of this "futurist" myth is that even as it continues to uphold the modern ideal of a general enlightenment and empowerment of the human subject, it does so by privileging certain individuals as being in advance of the general population—as already more enlightened and empowered. The problem here is not, however, *simply* a matter of a failure of "universal access," a failure to include everyone in the "High-Tech Enlightenment," the "High-Tech Future." The problem—as Horkheimer and Adorno, and many others who have followed them, have observed—is that this utopian "enlightenment" is premised on a notion of human mastery that inevitably involves a domination of "others." The mastery or empowerment of the human subject only makes sense so long as one has power over someone or something. That is to say, the idea of an active human subject (whether individual or collective) is founded on maintaining certain boundaries, a certain exclusion—which has, of course, served to justify the domination of various peoples who have been cast as less than full, active subjects: as inhuman, barbaric, animalistic, or childlike. Yet, even if all human beings are included as subjects, this notion of human mastery requires that the rest of the world be viewed as objectified fodder for human use, the stuff of domination. Thus, a certain repression is always inherent in this vision of a utopian, enlightened future; someone or something is always denied access.

Contesting this future, reimagining it, cannot, therefore, be figured *simply* in terms of "accessibility" or "inclusion." Lack of access will always remain a problem so long as we continue to view technology and the world in terms of instrumentality and human mastery, for the very idea of mastery is itself exclusionary. The problem is to imagine a "subject position" and a knowledge that are not based on an instrumental or technological mastery; the question is how to accept and live with "other" agencies, including the sublime, inhuman, and at times monstrous agency of the techno-cultural world, the techno-cultural unconscious—without, on the one hand, appropriating that agency to justify our own power over others or, on the other hand, feeling dominated or enslaved by it.

The representations of Butler, Haraway, and Gibson, among others, seem to suggest an increasing willingness to accept the uncontrollable, inhuman status of a techno-cultural unconscious, to accept the fact that, although human actions can affect it, they cannot simply control it.[47] This acceptance of the autonomy of techno-cultural processes implies a shift in the relation of human beings not only to technology, but to others. It implies a politics that is no longer based solely on the knowledge and actions of a "human subject." From this perspective, humanity can no longer be defined in opposition to its others, whether that otherness is technological, irrational, primitive, or simply monstrous. Instead, human beings begin to find themselves, like Haraway's cyborgs, in a much more mixed or hybrid subject position, which cannot easily be separated from the processes of the techno-cultural unconscious, or from others. Networked through techno-culture, their sense of identity need not rely on maintaining a rigid boundary between the "self" and others, whether technological or human. Indeed, their identity—no longer based on such originary myths as "humanity"—becomes relational, interactive, a matter of making connections, mutating and evolving, generating new codes and patterns from the fragments of the old. Neither masters nor slaves, subjects nor objects, they nevertheless have an agency, even if that agency—no longer simply human—cannot be divorced from the complex web of forces around them.

The consequences of techno-cultural processes are not inevitable, as if decreed by the blind gods of a mechanical fate. Like the Freudian unconscious, the realm of techno-culture is at once overdetermined and constantly in process; it is coded, but its codes are continually subject to mutation and rewriting. In this realm, nothing is inherently stable, secure, guaranteed. Such a realm is precisely the realm of politics, where futures are imagined, contested, and brought into being. If there is to be a techno-cultural politics that does not simply try to control the processes of techno-culture, it must imagine human beings as participants in the techno-cultural unconscious—riding its waves, attempting to navigate its currents, but also, by their actions, initiating unsettling new movements within it, generating new relations and processes, whose consequences often cannot be foreseen. Such a politics would itself be a complex, generative process in which fixed values and power relations would be unsecured, in which connection and interconnection with others would be "essential," in which hybridity and partiality would be valued over purity and wholeness.

Yet, as both Haraway and Gibson have suggested, the realm of technoculture is also a science-fictional realm, where small changes can generate profound and unpredictable mutations in the future. Indeed, the processes of the techno-cultural unconscious *are* the processes through which the future emerges. In such a realm, however, the future need not be simply "human," need not be predicated solely on the "utopian" politics of human enlightenment and empowerment; other futures are possible, imaginable.

To imagine our relations to the techno-cultural unconscious is to imagine our relations both to "others" and to these "other" futures. These "other" futures cannot be represented through rational analysis and prediction; they can only be imagined through a science-fictional process—an imaginative, aesthetic process that is similar to the "bringing forth" that Heidegger saw in the Greek *technē*. In imagining, as Butler, Haraway, and Gibson do, our own virtual gods and goddesses, our own alien and cyborg myths, to represent our relationship to technology, to the techno-cultural other, we are already participating in the ongoing processes and movements, in the science-fictional mutations and evolutions, in the high technē, through which other futures emerge, are brought to life, "brought forth."

Notes

Introduction

1. Jean Baudrillard, *For a Critique of the Political Economy of the Sign* (Saint Louis: Telos Press, 1981), p. 185.

2. Martin Heidegger, "The Question concerning Technology," in *The Question concerning Technology and Other Essays,* trans. William Lovitt (New York: Harper Torchbooks, 1977), p. 4; hereafter, *TQCT.*

3. In this book, I have consistently used the term *modernity* to refer to modernity in the broad sense—that is, to that shift in attitudes toward the world associated with the Renaissance—and the term *modernism* to refer to artistic modernism, beginning in the second half of the nineteenth century, but most commonly associated with the avant-garde artistic movements of the 1920s. *Modern* may refer to either, depending on the context, but often serves to suggest the continuities between the two.

4. On this point, cf. Bernard Stiegler's extended discussion and critique of Rousseau's notion of an originary, nontechnical human, whose existence would be prior to the "fall" into the "supplementarity" of technology, civilization, and history: "Technology and Anthropology," in *Technics and Time,* vol. 1, *The Fault of Epimetheus,* trans. Richard Beardsworth and George Collins (Stanford, Calif.: Stanford University Press, 1998), pp. 82–133. For Stiegler, there is no time (particular or general) prior to, or external to, technics.

5. Fredric Jameson, "The Cultural Logic of Late Capitalism," in *Postmodernism, or, The Cultural Logic of Late Capitalism* (Durham, N.C.: Duke University Press, 1991), p. 37. Page numbers for subsequent references to this work, as well as others, are given in the text.

6. James Botkin, Dan Dimancescu, and Ray Strata, with John McClellan, *Global Stakes: The Future of High Technology in America* (New York: Penguin Books, 1984), p. 20.

7. Samuel Weber, "Upsetting the Set Up: Remarks on Heidegger's Questing after Technics," *MLN* 104(5) (December 1989): 982.

8. Samuel Weber, "Theater, Technics, and Writing," *1–800* (fall 1989): 20 n. 8.

9. The constellation of meanings in this notion of "setting up" or "setting forth" is, for Heidegger, related to the whole family of verbs related to *stellen* (to set or place): for example, *aufstellen* (to set up), *bestellen* (to order, to set in order), *darstellen* (to present or exhibit), *herstellen* (to produce, to set here), *vorstellen* (to represent).

10. Stiegler, in fact, critiques Heidegger for failing to distinguish between "instruments" and "tools," citing the possibility of an artistic "instrument": "Not only does Heidegger think the instrument; he thinks *on the basis of it.* Yet he does not think it fully through: he fails to see in the instrument the originary and originarily deficient horizon of any discovery, including the unforeseen; he fails to see in the instrument what truly sets in play the temporality of being, what regarding access to the past and, therefore, to the future, is constituted through the instrument techno-logically, what through it constitutes the historical as such. He always thinks tools as (merely) useful and instruments (merely) *as* tools, and he is as a result incapable of thinking, for example, an artistic instrument as something that *orders* a world" (Stiegler, *The Fault of Epimetheus,* p. 245).

11. Weber, "Upsetting the Set Up," 987.

12. Le Corbusier, *Towards a New Architecture,* trans. Frederick Etchells (New York: Dover Publications, 1986), p. 95.

13. "What Does LEF Want?" *LEF,* no. 1 (March 1923); cited in Camilla Gray, *The Russian Experiment in Art, 1863–1922,* rev. ed. (New York: Thames & Hudson, 1986), p. 258.

14. Walter Benjamin, "Paris, Capital of the Nineteenth Century," in *Reflections,* trans. Edmund Jephcott (New York: Schocken Books, 1986), p. 161.

15. "To perceive the aura of an object we look at means to invest it with the ability to look at us in return" (Walter Benjamin, "On Some Motifs in Baudelaire," in *Illuminations,* trans. Harry Zohn [New York: Schocken Books, 1969], p. 188).

16. Peter Wollen, *Raiding the Icebox: Reflections on Twentieth-Century Culture* (Bloomington: Indiana University Press, 1993), p. 48.

17. Reyner Banham, *Theory and Design in the First Machine Age,* 2d ed. (Cambridge: MIT Press, 1980 [1960]).

18. Clarence Barnhart, Sol Steinmetz, and Robert K. Barnhart, *The Second Barnhart Dictionary of New English* (New York: Harper & Row, 1980).

19. Katherine Hayles has in fact argued that the emergence of "chaos theory" is linked to "information theories and technologies" that promote "the separation of information from meaning" (in *Chaos Bound: Orderly Disorder in*

Contemporary Literature and Science [Ithaca, N.Y.: Cornell University Press, 1990], p. 8).

20. Constance Penley and Andrew Ross, "Interview with Donna Haraway," in *Technoculture,* ed. Constance Penley and Andrew Ross (Minneapolis: University of Minnesota Press, 1991), p. 3. For Gibson's comments, see Kevin Kelly, "Cyberpunk Era: Interviews with William Gibson," *Whole Earth Review* 16 (summer 1989): 79. See also Sterling's "Preface" to *Mirrorshades: The Cyberpunk Anthology,* ed. Bruce Sterling (New York: Arbor House, 1986), p. ix.

21. Jacques Derrida, "Plato's Pharmacy," in *Dissemination,* trans. Barbara Johnson (Chicago: University of Chicago Press, 1981) p. 109; see esp. pp. 102–17.

22. Michael Schrage, "Software Bugs Poised to Bite as Phone System Expands," *Los Angeles Times,* October 5, 1989, business section, pp. 1, 7.

23. Mark Dery gives a rather caustic account of these and other cybercultural movements in his *Escape Velocity: Cyberculture at the End of the Century* (New York: Grove Press, 1996). For an insider's perspective on techno-paganism, see the special issue of the pagan zine *Green Egg* on techno-pagans: *Green Egg* 29(120) (August–September 1997). See also Erik Davis, "Technopagans," *Wired* 3.07 (July 1995): 126–33, 174–81.

24. William Gibson, *Count Zero* (New York: Ace Books, 1986).

25. Donna Haraway, "A Manifesto for Cyborgs: Science, Technology and Socialist Feminism in the 1980's," *Socialist Review* 80 (1985): 65–108; reprinted as "A Cyborg Manifesto: Science, Technology, and Socialist-Feminism in the Late Twentieth Century," in Donna J. Haraway, *Simians, Cyborgs, and Women: The Reinvention of Nature* (New York: Routledge, 1991), pp. 149–181.

26. Octavia Butler, "Bloodchild," in *Bloodchild and Other Stories* (New York: Seven Stories Press, 1996); *Clay's Ark* (New York: Warner Books, 1984). The "Xenogenesis" trilogy includes *Dawn* (New York: Popular Library, 1987), *Adulthood Rites* (New York: Popular Library, 1988), and *Imago* (New York: Popular Library, 1989).

27. Donna J. Haraway, "The Biopolitics of Postmodern Bodies: Constitutions of Self in Immune System Discourse," in *Simians, Cyborgs, and Women,* p. 227.

1. The Spirit of Utopia and the Birth of the Cinematic Machine

1. An apparently "premature" example of this "modern" birth process can be seen in the obstetric obsessions of Laurence Sterne's *Tristram Shandy,* where the intervention of the techniques and tools (e.g., the forceps) of obstetrics—the descriptions of which are frequently borrowed directly from scientific works of the time—affects not only Tristram's "wholeness" but the narrative's. Not only the birth of the character Tristram Shandy but that of *Tristram Shandy,* the novel, is "deformed" by the introduction of a technical element. Cf. Victor Shklovsky,

"Sterne's *Tristram Shandy*: Stylistic Commentary," in *Russian Formalist Criticism: Four Essays,* ed. Lee T. Lemon and Marion J. Reis (Lincoln: University of Nebraska Press, 1965), pp. 25–57.

2. Christian theology is certainly not immune to this tradition, particularly as it attempts to deal with science and mathematics in the early Renaissance. A tendency toward Platonic mathematical mysticism is particularly obvious in, for example, the works of Nicolas of Cusa and Marsilio Ficino. On the connection between god and science in Newton's work, see "God and Natural Philosophy," in *Newton's Philosophy of Nature: Selections from His Writings,* ed. H. S. Thayer (New York: Hafner Press, 1974), pp. 41–46.

3. Sigmund Freud, "The 'Uncanny,'" in *The Standard Edition of the Complete Psychological Works of Sigmund Freud,* vol. 17 (London: Hogarth Press, 1900), pp. 217–52.

4. See ibid., p. 240: "Our analysis of instances of the uncanny has led us back to the old, animistic conception of the universe. This was characterized by the idea that the world was peopled with the spirits of human beings; by the subject's narcissistic overvaluation of his own mental processes; by the belief in the omnipotence of thoughts and the technique of magic based on that belief; by the attribution to various outside persons and things of carefully graded magical powers, or 'mana'; as well as by all the other creations with the help of which man, in the unrestricted narcissism of that stage of development, strove to fend off the manifest prohibitions of reality." These laws of reality (which provide the basis for Freud's ethnocentrism here) are clearly the laws of scientific-technological, causal explanation and discourse. This discourse indeed represses (Freud, in his scientism, prefers "surmounts") the animistic or cosmological view of the world; it is, in fact, through the shift from a magical or theological to a scientific-technological discourse that techno-teleological modernity is born.

5. Lacan refers to this spirit as the *moi* or ego: "The *moi,* the ego, of modern man . . . has taken on its form in the dialectical impasse of the *belle âme* who does not recognize his very own *raison d'être* in the disorder that he denounces in the world." He also links this ego to modern science. See Jacques Lacan, "The Function and Field of Speech and Language in Psychoanalysis," in *Écrits: A Selection,* trans. Alan Sheridan (New York: W. W. Norton, 1977), pp. 70–71.

6. Walter Benjamin, "On Some Motifs in Baudelaire," *Illuminations,* trans. Harry Zohn (New York: Schocken Books, 1969), p. 188.

7. Walter Benjamin, "N [Theoretics of Knowledge, Theory of Progress]," *Philosophical Forum* 15 (1–2) (fall–winter 1983–84): 6–7 [N 2a, 1], [N 2a, 2].

8. The phrase is Baudelaire's; cited by Benjamin, "On Some Motifs in Baudelaire," p. 175.

9. Cf. Jacques Derrida, "Plato's Pharmacy," in *Dissemination,* trans. Barbara Johnson (Chicago: University of Chicago Press, 1981), esp. pp. 102–17.

10. Walter Benjamin, "Paris, Capital of the Nineteenth Century," in *Reflections,* trans. Edmund Jephcott (New York: Schocken Books, 1986), p. 161.

11. Walter Benjamin, "The Work of Art in the Age of Mechanical Reproduction," in *Illuminations,* p. 244 n. 7; hereafter, "WAAMR."

12. Noël Burch, "Charles Baudelaire versus Doctor Frankenstein," *Afterimage,* nos. 8–9 (spring 1981): 4–21.

13. Benjamin, "On Some Motifs in Baudelaire," p. 186. See Charles Baudelaire, "The Salon of 1859," in *Art in Paris, 1845–1862: Salons and Other Exhibitions,* trans. and ed. Jonathan Mayne (London: Phaidon Publishers, 1965), p. 154: "Let it [photography] rescue from oblivion those tumbling ruins, those books, prints and manuscripts which time is devouring, precious things whose form is dissolving and which demand a place in the archives of our memory—it will be thanked and applauded. But if it be allowed to encroach upon the domain of the impalpable and the imaginary, upon anything whose value depends solely upon the addition of something of a man's soul, then it will be so much the worse for us!"

14. Panofsky: "anatomy as a science (and this applies to all the other observational or descriptive disciplines) was simply not possible without a method of preserving observations in graphic records, complete and accurate in three dimensions. . . . It is no exaggeration to say that in the history of modern science the advent of perspective marked the beginning of a first period; the invention of the telescope and the microscope that of a second; the discovery of photography that of a third: in the observational or descriptive sciences illustration is not so much the elucidation of a statement as a statement in itself" ("Artist, Scientist, Genius: Notes on the 'Renaissance-Dämmerung,'" in *The Renaissance: Six Essays* [New York: Harper Torchbooks, 1962], p. 147).

15. E. J. Marey, preface to Eugène Trutat, *La photographie animée* (Paris: Gauthiers-Villars, 1899); cited in Burch, p. 13.

16. Georges Sadoul, *Louis Lumière* (Paris: Éditions Seghers, 1964), p. 107; cited in Burch, p. 19.

17. There is, in my view, some reason to doubt Burch's view of the Lumière films as somehow "scientific" in their approach. There seems little effort in them to analyze the recorded data, to make use of it as, for example, Marey did. The polycentrism that Burch observes in these films is, indeed, quite remarkable, but it does not seem particularly "scientific." On the issue of the relation of cinema and montage to science, see my discussion of the filmic avant-garde in chapter 3.

18. Benjamin calls this memory—to which Baudelaire assigns photography—"volitional, discursive memory" ("On Some Motifs in Baudelaire," p. 186).

19. Benjamin, "The Work of Art in the Age of Mechanical Reproduction," p. 236.

20. James Pettifer provides a working definition of the standard German usage of *Montage.* It is used to refer to "a work whose openness and heterogeneity is characterized by assembling (simultaneously or successively) a series of strongly homogeneous units within a deliberately incongruous frame (e.g., Heartfield's photomontages, Piscator's stagings, Eisenstein's films, particularly *October* and *The General Line*) ("Against the Stream: *Kuhle Wampe,*" *Screen* 15[2]: 64 n. 10).

21. Burch cites the following reviews of the first showing of the Cinematograph as indications of its annexation "to the 'Frankensteinian' ideological field": from *Le Radical*: "Whatever the scene thus filmed and however great the number of individuals taken unawares in their daily lives, you see them life size, in color *[sic],* in perspective, with distant skies, houses, the complete illusion of real life.... Already speech had been recorded and reproduced. You could, for instance, see your loved ones again long after they are lost to you." From *La Poste*: "When these cameras become available to the public, when all are able to photograph their dear ones, no longer merely in immobile form but in movement, in action, with their familiar gestures, with speech on their lips, death will no longer be final" (pp. 20–21).

22. Benjamin's quotations from Gance and Séverin-Mars, respectively, are as follows: "Here, by a remarkable regression, we have come back to the level of expression of the Egyptians.... Pictorial language has not yet matured because our eyes have not yet adjusted to it. There is as yet insufficient respect for, insufficient cult of, what it expresses." "What art has been granted a dream more poetical and more real at the same time! Approached in this fashion the film might represent an incomparable means of expression. Only the most high-minded persons, in the most perfect and mysterious moments of their lives, should be allowed to enter its ambience." The source for both quotations is Abel Gance, "Le Temps de l'image est venu," in *L'Art cinématographique,* vol. 2 (Paris: Librairie Félix Alcan, 1926), pp. 100–101; cited in Benjamin, "The Work of Art in the Age of Mechanical Reproduction," p. 227.

23. André Bazin, "The Ontology of the Photographic Image," in *What Is Cinema?* vol. 1, trans. Hugh Gray (Berkeley: University of California Press, 1967), p. 9.

24. André Bazin, "The Myth of Total Cinema," in *What Is Cinema?* vol. 1, pp. 17–22.

25. Shelley's Frankenstein is not actually referred to as "Doctor"; he does, however, commit himself to the study of "natural philosophy," which includes all the branches of the physical sciences. He is certainly, therefore, a scientist.

Moreover, his interest in the biological sciences makes the later, movie attribution of "Dr." appropriate.

26. Mary Shelley, *Frankenstein: Or, The Modern Prometheus* (New York: Penguin Books, 1963), chap. 2, p. 37.

27. Ibid., pp. 39–40. The passage continues: "Nor were these my only visions. The raising of ghosts or devils was a promise liberally accorded by my favourite authors, the fulfilment of which I most eagerly sought; and if my incantations were always unsuccessful, I attributed the failure rather to my own inexperience and mistake than to a want of skill or fidelity in my instructors."

28. Benjamin, "On Some Motifs in Baudelaire," p. 186.

29. Paul de Man, "Literary History and Literary Modernity," in *Blindness and Insight: Essays in the Rhetoric of Contemporary Criticism,* 2d rev. ed. (Minneapolis: University of Minnesota Press, 1983), pp. 157–58.

30. Charles Baudelaire, "The Painter of Modern Life," in *The Painter of Modern Life and Other Essays,* trans. and ed. Jonathan Mayne (London: Phaidon Press, 1964), p. 17 (translation altered slightly).

31. As Matei Calinescu notes, for Baudelaire, "The task of the artist is akin to the alchemical one of extracting gold from mud or—if we translate this typically Baudelairean metaphor—to reveal the poetry hidden behind the most horrifying contrasts of social modernity" (*Five Faces of Modernity* [Durham, N.C.: Duke University Press, 1987], p. 54).

32. Ibid., p. 57.

33. Michel Carrouges, *Les machines célibataires* (Paris: Arcanes, 1954).

34. See Harald Szeeman, ed., *Le macchine celibi/The Bachelor Machines* (New York: Rizzoli, 1975), with essays by Carrouges, Lyotard, Schwarz, and others. See also, Gilles Deleuze and Félix Guattari, *Anti-Oedipus: Capitalism and Schizophrenia,* trans. Robert Hurley, Mark Seem, and Helen R. Lane (Minneapolis: University of Minnesota Press, 1983), esp. pp. 17–18; Jean-François Lyotard, *Dérive à partir de Marx et Freud* (Paris: Union Générale d'Éditions, 10/18, 1973); Michel de Certeau, "The Arts of Dying: Celibatory Machines," in *Heterologies: Discourse on the Other,* trans. Brian Massumi (Minneapolis: University of Minnesota Press, 1986), pp. 156–67; Arturo Schwarz, *The Alchemist Stripped Bare in the Bachelor, Even* (New York: Museum of Modern Art, 1973), and *The Complete Works of Marcel Duchamp* (New York: Abrams, 1969).

35. Michel Carrouges, "Directions for Use," in Szeeman, *Le macchine celibi,* p. 21.

36. Arturo Schwarz, "The Alchemical Bachelor Machine," in Szeeman, *Le macchine celibi,* p. 156.

37. In alchemy, gold is the sublimated form of the base metal. The play on

aura–aurea is a false etymology (*aura* is from the Greek *aer*: air), but the two are related in *aureole.*

38. Constance Penley gives a succinct summary of this "fantasy": "Characteristically, the bachelor machine is a closed, self-sufficient system. Its common themes include frictionless, sometimes perpetual motion, an ideal time and the magical possibility of its reversal (the Time Machine is a bachelor machine), electrification, voyeurism and masturbatory eroticism, the dream of the mechanical reproduction of art, and artificial birth or reanimation. But no matter how complicated the machine becomes, the control over the sum of its parts rests with a knowing producer who therefore submits to a fantasy of closure, perfectibility and mastery" ("Feminism, Film Theory and the Bachelor Machines," *m/f* 10 [1985]: 39).

39. Marcel Duchamp, *Notes and Projects for the Large Glass* (London: Thames & Hudson, 1969), p. 23.

40. Villiers de l'Isle-Adam, *Tomorrow's Eve,* trans. Robert Martin Adams (Urbana: University of Illinois Press, 1982), p. 3.

41. The Artist or the Artifex are the names by which alchemists referred to themselves.

42. Villiers, *Tomorrow's Eve,* p. 7 (translation altered slightly).

43. Benjamin: "only a redeemed mankind receives the fullness of its past—which is to say, only for redeemed mankind has its past become citable in all its moments" ("Theses on the Philosophy of History," in *Illuminations,* p. 254).

44. Raymond Bellour, "Ideal Hadaly," trans. Stanley E. Gray, *Camera Obscura* 15 (1986): 113.

45. Villiers: "[I]n him who reflects it, the living idea of God appears only to the extent that the faith of the viewer is able to evoke it. Like every other thought, God can exist in the individual only according to the capacity of the individual. No one knows where illusion begins or what reality consists of. Thus, God being the most sublime of conceptions, and all conceptions existing only according to the particular spirit and the intellectual vision of the seer, it follows that the man who dismisses the idea of God from his thoughts does nothing but deliberately decapitate his own mind" (p. 24).

46. As will become apparent, the ability to provide a blank screen or echo chamber for Ewald's desires is not entirely the result of the technological body of Hadaly.

47. This "secret" also explains Edison's surprise at discovering that he can no longer contact Sowana after Hadaly's departure.

48. The electrical "connection" between the material, technological world of Edison and the eternal, spiritual world of Sowana runs throughout *L'Ève future*

and is truly the "wiring" that holds its disparate parts together. And just as this electrical metaphor will animate *L'Ève future,* it will also be conducted to those works influenced by Villiers's novel, notably, Whale's *Bride of Frankenstein* and Lang's *Metropolis.*

49. Annette Michelson, in her article on *L'Ève future,* refers to this action of fragmentation and sublimation as "that proleptic movement between analysis and synthesis which will accelerate and crystallize around the female body in an ultimate, fantasmatic mode of representation as cinema" ("On the Eve of the Future: The Reasonable Facsimile and the Philosophical Toy," *October* 29 [summer 1984]: 20).

50. For suggesting this connection, and for its excellent summary of the bachelor machine phenomenon, I owe a large debt of gratitude to David J. Russell's conference paper, "Bachelor Machine: Black Maria/False Maria" (presented at the 1989 Society for Cinema Studies Conference).

51. In fact, Sadoul uses precisely these terms in describing the "birth" of narrative film in the Black Maria: "But if a dancer who performed in front of the black background of the Black Maria produced a quasi-abstract silhouette, in the sketches composed by Dickson, there were enough properties used for a *birth* of decor *(décor en naisse),* which created atmosphere and gave these films a much greater impression of reality than the nearly abstract schematization of those other films. . . . In certain films, moreover, he conceived an *embryo* of a scenario, which he used in short, thirty-second dramas and comedies" (Georges Sadoul, *Histoire générale du cinéma,* vol. 1, *L'invention du Cinéma, 1832–1897* [Paris: Denoël, 1948], pp. 246–48; my translation, emphasis added).

52. André Bazin, "The Virtues and Limitations of Montage," in *What Is Cinema?* vol. 1, p. 47.

53. André Bazin, "In Defense of Rossellini," in *What Is Cinema?* vol. 2, trans. Hugh Gray (Berkeley: University of California Press, 1971), pp. 97–98.

54. Both Villiers and Bazin were Roman Catholics. Bazin's work, as Hugh Gray has noted in his introduction to the second volume of *What Is Cinema?,* shows the influence of Teilhard de Chardin. In several passages, Bazin speaks of the neorealists' love of the world as the unifying force that lets them see the world as whole. A more thorough analysis of this facet of Bazin's thought—which is merely suggested in this chapter—would certainly seem worthwhile.

55. See Bazin, "The Virtues and Limitations of Montage," pp. 41–52, and also the footnote on page 19 of "The Myth of Total Cinema."

56. Benjamin, "On Some Motifs in Baudelaire," pp. 187–88.

57. André Bazin, "Theater and Cinema, Part Two," in *What Is Cinema?* vol. 1, p. 97.

2. The Mediation of Technology and Gender

1. Andreas Huyssen, "The Vamp and the Machine: Fritz Lang's *Metropolis*," in *After the Great Divide: Modernism, Mass Culture, Postmodernism* (Bloomington: Indiana University Press, 1986), p. 67.

2. Andreas Huyssen, "The Hidden Dialectic: Avantgarde—Technology—Mass Culture," in *After the Great Divide*, p. 10.

3. Lang claimed that the idea for *Metropolis* originated while viewing the New York skyline during his 1924 trip to the United States. His wife, Thea von Harbou, then wrote a novel based on this idea, which the two adapted into a scenario. For its 1927 American release, the film was drastically reedited, slicing approximately seven reels from its original seventeen. The subsequent German rerelease of the film followed the cuts of the American version fairly closely, with the result that Lang's original version has, as Enno Patalas notes, "been thoroughly and irreparably destroyed, as few other films have been." Efforts to reconstruct the missing footage and intertitles have, however, helped to restore sense to some of the film's scenes, particularly those between Fredersen and Rotwang (on this point, see note 8 on the excised "Hel subplot"). Patalas's reconstructive work has been especially helpful in this regard; see, for example, his "*Metropolis*, Scene 103," trans. Miriam Hansen, *Camera Obscura* 15 (fall 1986): 165–72, as well as the version of *Metropolis* released by Giorgio Moroder in 1984, which, largely based on Patalas's work, restores several important "lost scenes" to the film.

It is worth noting, however, that the subplots and motivations restored to the film by these reconstructions can also be found in Thea von Harbou's novel, *Metropolis* (Boston: Gregg Press, 1975; rpt. of 1929 edition published by The Reader's Library), which I have found extremely helpful to my interpretation of the film. Throughout this chapter, then, I have drawn material not only from the edited and the reconstructed versions of the film, but also from the novel. Where my reading depends on material drawn from the novel, I have attempted to designate the source as clearly as possible.

4. Frederick W. Taylor's *Principles of Scientific Management* was published in 1911. Henry Ford's first assembly line started not long after. For an account of the importance of Taylor's and Ford's ideas for cinema, and a host of other interesting ideas, see Peter Wollen, "Cinema/Americanism/The Robot," *New Formations* 8 (summer 1989): 7–34.

5. The workers, as Maria notes in the parable of the Tower of Babel, know "nothing of the dream of those who planned the tower."

6. This triadic structure has deep roots in German philosophy; it should recall not only Hegel, but Kant's three Critiques.

7. Johann Wolfgang von Goethe, *Faust II*, ll. 12110–111, in *Goethe's Faust*, trans. Walter Kaufman (New York: Doubleday, 1961). This connection is support-

ed by the Faustian overtones that appear throughout *Metropolis*—particularly in relation to the Fredersen character—and that are considerably clearer in Thea von Harbou's novel, as they apparently were in the original version of the film.

8. The "Hel subplot" was removed by the film's American editors. The justification for removing the scene that explains the relationship between Fredersen, Rotwang, and Hel is stated in the following, often-cited, passage: "[The scene] showed a very beautiful statue of a woman's head, and on the base was her name—and that name was 'Hel.' Now the German word for 'hell' is 'Hölle,' so they were quiet *[sic]* innocent of the fact that this name would create a guffaw in an English-speaking country. So it was necessary to cut this beautiful bit out of the picture, and a certain motive which it represented had to be replaced by another" (Randolph Bartlett, "German Film Revision Upheld as Needed Here," *New York Times*, March 13, 1927). The relationship between Fredersen, Rotwang, and Hel has, however, been made clear by Patalas's reconstruction in "*Metropolis*, Scene 103," and can also be seen in the Moroder version of the film, as well as in von Harbou's novel. Hel, although loved by Rotwang, had succumbed to Fredersen's will and gone away with him. She died giving birth to Freder, but the novel suggests that it was actually Fredersen's excessive love and her own guilt over leaving Rotwang that killed her. This information obviously gives the motivation for many of Rotwang's seemingly pointless actions in the film, including his actions at the end of the film where, thinking he has died, he goes in search of Hel and, mistaking Maria for her, attempts to carry Maria off.

9. Roger Dadoun, "*Metropolis*: Mother-City—'Mittler'—Hitler," trans. Arthur Goldhammer, *Camera Obscura* 15 (fall 1986): 149.

10. Siegfried Kracauer, *From Caligari to Hitler: A Psychological History of the German Film* (Princeton, N.J.: Princeton University Press, 1947), p. 107.

11. Klaus Theweleit, *Male Fantasies*, 2 vols., trans. Stephen Conway, Erica Carter, and Chris Turner (Minneapolis: University of Minnesota Press, 1987, 1989).

12. Although the scope of this chapter does not allow a detailed treatment of this point, it would be remiss not to mention the extent to which this sense of a "split" or divided modernity is related to, and indeed dependent on, the German experience of World War I. Theweleit's work, of course, makes this experience a central point in his analysis of German culture between the wars.

13. Cf. Thomas Elsaesser's "Social Mobility and the Fantastic: German Silent Cinema," *Wide Angle* 5(2) (1982): 14–25, where Elsaesser reads the figure of "the Double" in German silent cinema both as a displacement of the "fear of proletarianization" (or industrialization) and as the return of a repressed social history "in the form of the uncanny and fantastic."

14. On Metropolis as mother city, see Dadoun, *Metropolis*. Although Dadoun

suggests a link between the notion of the mother city and Hitler, he omits the connection between mother city and Hitler's references to Germany as "the Motherland."

15. Kenneth Frampton, *Modern Architecture: A Critical History*, rev. and enl. ed. (London: Thames & Hudson, 1985), p. 116. This division between the typical and the "will to form" echoes the two antagonistic aspects of William Worringer's cultural model, as presented in his widely read *Abstraction and Empathy* (*Abstraktion und Einfühlung*, 1908). Against the tendency toward abstract types, Worringer posed, and popularized, a notion of Empathy as the projection of an individual creative will into the art object, thus conflating psychologist Theodor Lipps's idea of *Einfühlung* with art historian Alois Riegl's notion of *Kunstwollen*.

16. Von Harbou, *Metropolis*, p. 154. Interestingly, in the novel, Fredersen has had the house transplanted, garden and all, to the top of one of the city's skyscrapers.

17. Sigmund Freud, "The 'Uncanny,'" in *The Standard Edition of the Complete Psychological Works of Sigmund Freud*, vol. 17 (London: Hogarth Press, 1900), pp. 217–52.

18. Freud notes, "It often happens that male patients declare that they feel there is something uncanny about the female genital organs. This *unheimlich* place, however, is the entrance to the former *Heim* [home] of all human beings" (ibid., p. 245).

19. Max Weber, *The Theory of Social and Economic Organization*, trans. A. M. Henderson and Talcott Parsons (New York: Oxford University Press, 1947), p. 185.

20. Frampton, *Modern Architecture*, p. 216.

21. From Hitler's speech at the closing rally of the 1934 Nuremberg rally, recorded in Riefenstahl's *Triumph of the Will*.

22. Theodor W. Adorno, "Freudian Theory and the Pattern of Fascist Propaganda," in *The Essential Frankfurt School Reader*, ed. Andrew Arato and Eike Gebhardt (New York: Continuum, 1982), p. 123.

23. Walter C. Langer, *The Mind of Adolf Hitler* (New York: Basic Books, 1972), p. 153.

24. Anton Kaes, *From* Hitler *to* Heimat*: The Return of History as Film* (Cambridge: Harvard University Press, 1989), p. 147.

25. Friedrich Schiller, *On the Aesthetic Education of Man*, trans. Reginald Snell (New York: Unger, 1965), p. 40.

26. Peter Bürger, *Theory of the Avant-Garde*, trans. Michael Shaw (Minneapolis: University of Minnesota Press, 1984), p. 45.

27. G. W. F. Hegel, *Aesthetics: Lectures on Fine Arts*, trans. T. M. Knox (Oxford: Clarendon Press, 1975), p. 38.

28. Hegel expresses this view of "art as organic" quite clearly while summa-

rizing Kant's aesthetics, with which on this point he obviously agrees: "The beautiful . . . exists as purposeful in itself, without means and end showing themselves separated as different aspects of it. The purpose of the limbs, for example, of an organism is the life which exists as actual in the limbs themselves; separated they cease to be limbs. For in a living thing purpose and the material for its realization are so directly united that it exists only in so far as its purpose dwells in it" (ibid., p. 59).

29. Walter Benjamin, "On Some Motifs in Baudelaire," in *Illuminations,* trans. Harry Zohn (New York: Schocken Books, 1969), p. 188.

30. Charles Baudelaire, "The Painter of Modern Life," in *The Painter of Modern Life and Other Essays,* trans. and ed. Jonathan Mayne (London: Phaidon Press, 1964), pp. 12–13.

31. The concern with abstraction in these movements was based less on notions of technological rationality or functionalism than on the idea of abstract, geometric forms that they conceived in Neoplatonic terms, as universal and eternal. Neo-Plasticism and early de Stijl were also heavily influenced by Theosophy.

32. Bürger, *Theory of the Avant-Garde,* p. 49.

33. For an interesting reading of the Gothic cathedral, which echoes and amplifies many of the ideas expressed here, see William Worringer's *Form in Gothic* (New York: Schocken Books, 1964), a work that, as Worringer notes, is a kind of sequel to his *Abstraction and Empathy* (on which, see note 15).

34. Walter Gropius, "Program of the Staatliche Bauhaus in Weimar," in *The Bauhaus: Weimar, Dessau, Berlin, Chicago,* ed. Hans M. Wingler, trans. Wolfgang Jabs and Basil Gilbert (Cambridge: MIT Press, 1969), p. 31.

35. Ernst Bloch, "Building in Empty Spaces," in *The Utopian Function of Art and Literature,* trans. Jack Zipes and Frank Mecklenburg (Cambridge: MIT Press, 1988), p. 195.

36. Cf. Joseph Goebbels's description of the "expression" of this cathedral of the future in his novel *Michael*: "As the man above us spoke, stone rolled upon stone, building the *cathedral of the future.* For years, something had been living inside me: it now took shape and became palpable.

"Revelation! Revelation!

"In the midst of ruin, a single man had dared to stand and raise the banner high" (*Michael; ein deutsches Schicksal in Tagebuchblättern* [Munich: Verlag Eher, 1942, 1929], p. 102; my translation; emphasis added).

37. Ibid., p. 21.

38. It is worth noting that in his 1927 essay "The Mass Ornament" (trans. Barbara Correll and Jack Zipes, *New German Critique* 5 [spring 1975]: 67–76) Kracauer argues that capitalism, with which he associates the mass ornament, "does not rationalize too much but *too little.*" Thus, "the weight granted to reason

in the ornament is . . . an illusion," an abstract, empty, formal rationality that allows "uncontrolled nature [to grow] prodigiously under the pretense of rational expression." For Kracauer, then, only a "true reason," one that is capable of "encompassing human beings," could overcome the formal rationality of the mass ornament. This "human reason" would, then, allow "ornaments" based not on the rationalized form of "the masses," but in the "organic life" of "the people." One cannot help but be struck here by the similarity between Kracauer's notion of a human reason—which would apparently reintegrate (synthesize?) an empty, modern rationalization with organic, human concerns—and the desire for a mediation of rationality and nature, the technological and the organic, in *Metropolis* and Nazism. This is not, of course, to conflate Kracauer's thinking with Nazi ideology, but merely to suggest that what blinds Kracauer to the ideological appeal of *Metropolis* and Nazism is precisely the similarity of that appeal (with its basis in a desire for mediation) to his own, clearly Hegelian, point of view.

39. This "truth" is, of course, ideological, as is the desire for wholeness that is its basis.

40. Huyssen, *After the Great Divide,* p. 73.

41. Speer's story of the background of the "cathedral of light" can be found in his *Inside the Third Reich,* trans. Richard and Clara Winston (New York: Macmillan, 1970), pp. 58–59.

42. This notion of a "connection" to the cosmos, to an eternal, celestial realm of the spirit, recurs constantly in Nazi symbology. It can be seen in, for example, the architectural eagle that hovers over Hitler in Speer's "cathedral of light," in the often-noted aerial shots of Hitler's plane at the beginning of *Triumph of the Will,* as well as in the film's final shot, where the men of the S.A. (the name itself *[Sturmabteilung]* alluding to the heavens) appear to march—in formation, as a mass ornament—into the clouds. It is, in fact, worth comparing this shot to the final scene of *Metropolis*: in both, the mass ornament is not portrayed as a purely abstract, mechanical form, but becomes a form reinfused with, a "technology" reanimated by, a spiritual aspect, a soul or aura. It thus becomes a "living," aesthetic whole.

43. This desire for an aesthetic, and indeed cinematic, mediation of the ancient and the modern, the past and the future, can also be seen in the cinema of the Third Reich as a whole. As Peter Reichel has noted, Nazi audiences were fascinated not only with historical films centered on famous German leaders, scientists, and artists of the past, but also with the future, with science-fiction films, many of which show the influence of *Metropolis.* See Peter Reichel, "Das Medium, das die Herzen erobert: Der Film," in his *Der schöne Schein des Dritten Reiches: Faszination und Gewalt des Faschismus* (Munich: Hanser, 1991).

44. Hans Jürgen Syberberg, "Hans Jürgen Syberberg—A Cycle Concluded," *Filmex Society Newsletter* 7 (June 1979): 3.

45. Gilles Deleuze, *Cinema 2: The Time-Image,* trans. Hugh Tomlinson and Robert Galeta (Minneapolis: University of Minnesota Press, 1989), p. 264.

46. Kaes, *From* Hitler *to* Heimat, p. 4.

47. The extent to which modernism is "haunted" by this "spirit" of aesthetic wholeness can be seen in the repeatedly voiced desire for a "total work of art," from Mallarmé's Total Book to Wagner's *Gesamtkunstwerk* to Bazin's "Myth of Total Cinema," as well as in the "alliance of the arts under the wing of a great architecture" announced in the manifesto of *Arbeitsrat für Kunst* and repeated in Gropius's Bauhaus manifesto.

48. In contrast to the Nazi view, see Jacques Lacan, "Aggressivity in Psychoanalysis," *Écrits: A Selection,* trans. Alan Sheridan (New York: W. W. Norton, 1977), pp. 8–29, where he notes the connection of aggressivity to an imaginary wholeness. For Lacan, moreover, it is precisely through Oedipal identification that this aggressivity is to be "transcended."

49. Richard Barsam, *Film Guide to* Triumph of the Will (Bloomington: Indiana University Press, 1975), p. 67.

3. The Avant-Garde Technē and the Myth of Functional Form

1. Charles Russell, *Poets, Prophets, and Revolutionaries: The Literary Avant-Garde from Rimbaud through Postmodernism* (Oxford: Oxford University Press, 1985), pp. 13–14.

2. Hugh Kenner, *The Mechanic Muse* (Oxford: Oxford University Press, 1987), p. 11.

3. Renato Poggioli, *The Theory of the Avant-Garde,* trans. Gerald Fitzgerald (Cambridge: Harvard University Press, 1968), p. 138.

4. Russell, *Poets, Prophets, and Revolutionaries,* p. 16. Russell's definition of the avant-garde generalizes Peter Bürger's definition of a socially activist "historical avant-garde" to include both Rimbaud and the "postmodernist avant-garde" and therefore conflicts with Bürger's historical argument that Futurism and the avant-gardes of the 1920s were a response to aestheticism that failed and died out. Bürger's argument, however, is itself belied by the historical presence of William Morris and the English Arts and Crafts movement, for whom the integration of art in the life-world was a major concern. Given the often-noted influence of this movement on modern art and architecture (particularly in Germany), this seems a curious omission.

5. Peter Bürger, *Theory of the Avant-Garde,* trans. Michael Shaw (Minneapolis: University of Minnesota Press, 1984).

6. Andreas Huyssen, "The Hidden Dialectic: Avantgarde—Technology—Mass Culture," in *After the Great Divide: Modernism, Mass Culture, Postmodernism* (Bloomington: Indiana University Press, 1986), p. 9.

7. Ibid., p. 11. The full quotation is given in the epigraph to this chapter.

8. The historical avant-garde is not, of course, a monolithic structure, nor can its relationship with high modernism be seen entirely as a matter of rupture. Indeed, in movements such as German Expressionism, just as in high modernism, a highly aestheticist attitude—a desire to subordinate technology, and technique, to an aesthetic or spiritual ideal—is evident (on this point, see chapter 2). At the other end of the spectrum, various members of, for example, the Productivist movement attempt to take a stance so strongly technicist that at times they eschew the aesthetic altogether. It could also be argued that in certain strains of Dadaism, both art and technology are abjured.

9. Address at the First International Congress of Architects, cited in Sigfried Giedion, *Space, Time and Architecture,* 5th rev. ed. (Cambridge: Harvard University Press, 1967), p. 216.

10. Walter Benjamin, "Paris, Capital of the Nineteenth Century," in *Reflections,* trans. Edmund Jephcott (New York: Schocken Books, 1986), p. 147.

11. Kenneth Frampton, *Modern Architecture: A Critical History,* rev. and enl. ed. (New York: Thames & Hudson, 1985), p. 34.

12. David Landes, *The Unbound Prometheus: Technological Change and Industrial Development from 1750 to the Present* (Cambridge: Cambridge University Press, 1969), p. 307.

13. Le Corbusier, *Towards a New Architecture,* trans. Frederick Etchells (New York: Dover Publications, 1986), p. 95; and "What Does LEF Want?" *LEF,* no. 1 (March 1923); cited in Camilla Gray, *The Russian Experiment in Art, 1863–1922,* rev. ed. (New York: Thames & Hudson, 1986), p. 258.

14. Nikolaus Pevsner, *Pioneers of Modern Design: From William Morris to Walter Gropius,* 3d rev. ed. (New York: Pelican Books, 1975; originally published as *Pioneers of the Modern Movement* by Faber and Faber, 1936), p. 26.

15. Adolf Loos, "Architektur," in *Trotzdem* (Innsbruck: Brenner-Verlag, 1931); cited in Frampton, *Modern Architecture,* p. 92.

16. Benjamin, "Paris, Capital of the Nineteenth Century," p. 149.

17. Stepanova's and Popova's statements are taken from the catalog of the September 1921 exhibition "5 × 5 = 25," the "last stand of easel painting" in Russia; cited in Gray, *The Russian Experiment in Art,* pp. 250–51.

18. Charles Caldwell, "Thoughts on the Moral and Other Indirect Influences of Rail-roads," *New England Magazine* 2 (April 1832); cited in Leo Marx, *The Machine in the Garden: Technology and the Pastoral Idea in America* (Oxford: Oxford University Press, 1964), p. 195.

19. Thomas Ewbank, U.S. Commissioner of Patents, quoted in *Scientific American* 5 (February 1850): 173; cited in Marx, *The Machine in the Garden,* p. 203.

20. Lothar Bucher, *Kulturhistoriche Skizzen aus der Industrie-ausstellung aller*

Völker (Frankfurt, 1851), p. 174; cited in Gidieon, *Space, Time and Architecture,* p. 253.

21. Théophile Gautier, in the newspaper *La Presse,* 1850; cited in Gidieon, *Space, Time and Architecture,* p. 215.

22. Reyner Banham, *Theory and Design in the First Machine Age,* 2d ed. (Cambridge: MIT Press, 1980 [1960]), p. 153.

23. Piet Mondrian, "de nieuwe beelding in de schilderkunst," *De Stijl* (Amsterdam) 1(1) (1917); cited in ibid., p. 150. A translation of the first section of this essay, taken from Michel Seuphor, *Piet Mondrian, Life and Work* (New York: Abrams, n.d.), appears in *Theories of Modern Art,* ed. Herschel B. Chipp (Berkeley: University of California Press, 1968), pp. 321–23, where it is erroneously dated 1919.

24. As John Willett notes, *nieuwe beelding* is more correctly translated as "new formation" or "new shaping" (John Willett, *Art and Politics in the Weimar Period: The New Sobriety, 1917–1933* [New York: Pantheon Books, 1978], p. 32).

25. Mondrian, cited in Banham, *Theory and Design in the First Machine Age,* p. 152.

26. Cited in ibid., p. 151.

27. Cited in Willett, *Art and Politics in the Weimar Period,* p. 32.

28. Kazimir Malevich, "From Cubism and Futurism to Suprematism: The New Painterly Realism" (Petrograd, 1915), in *Russian Art of the Avant-Garde,* 2d rev. ed., ed. and trans. John E. Bowlt (London: Thames & Hudson, 1988), p. 126.

29. For example, Malevich, "From Cubism and Futurism to Suprematism," p. 128:

"The art of utilitarian reason has a definite purpose.

But intuitive creation does not have a utilitarian purpose. Hitherto we have had no such manifestation of intuition in art.

Intuitive form should arise out of nothing.

Just as reason, creating things for everyday life, extracts them from nothing and perfects them.

Thus the forms of utilitarian reason are superior to any depictions in pictures.

They are superior because they are alive and have proceeded from material that has been given a new form for the new life. . . .

Here is a miracle.

There should be a miracle in the creation of art, as well."

30. Kazimir Malevich, "From *On New Systems in Art*" (Vitebsk, 1919), in *The Avant-Garde in Russia, 1910–1930,* ed. Stephanie Barron and Maurice Tuchman (Los Angeles: Los Angeles County Museum of Art, 1980), p. 201.

31. Vladimir Tatlin, "The Work Ahead of Us" (Moscow, January 1921), in *Russian Art of the Avant-Garde,* pp. 206–7.

32. Ilia Chashnick, "From 'A Department of Architecture and Technology,'" (Vitebsk, 1921), in *The Avant-Garde in Russia, 1910–1930,* p. 141.

33. Osip Brik, "From Pictures to Textile Prints" *LEF,* no. 2 (Moscow, 1924), in *Russian Art of the Avant-Garde,* p. 249.

34. Nikolai Punin, "Cycle of Lectures" (Petrograd, 1919), in *Russian Art of the Avant-Garde,* p. 174.

35. The formula is given by John Bowlt in his introduction to Punin's essay: $S(Pi + Pii + Piii + \ldots P)Y = T$, where S equals the sum of the principles (P), Y equals intuition, and T equals artistic creation (ibid., p. 171).

36. Liubov Popova, "On Exact Criteria, Ballet Numbers, Deck Equipment on Battleships, Picasso's Latest Portraits, and the Observation Tower at the School of Military Camouflage in Kuntsevo" (Moscow, 1922), in *The Avant-Garde in Russia, 1910–1930,* p. 222.

37. Cited in Dawn Ades, *Photomontage,* rev. ed. (London: Thames & Hudson, 1986), p. 13.

38. Raoul Hausmann, *Courrier Dada* (Paris: Le Terrain Vague, 1958), p. 42.

39. Gustav Klutsis, "Photomontage as a New Kind of Art of Agitation" (Moscow, 1931), cited in Szymon Bojko, *New Graphic Design in Revolutionary Russia,* trans. Robert Strybel and Lech Zembrzuski (New York: Praeger, 1972), p. 21.

40. Sergei Eisenstein, *Notebooks from 1919,* cited in Jacques Aumont, *Montage Eisenstein,* trans. Lee Hildreth, Constance Penley, and Andrew Ross (Bloomington: Indiana University Press, 1987), p. 150.

41. Sergei Eisenstein, "How I Became a Film Director," in *Notes of a Film Director,* ed. R. Yurenev, trans. X. Danko (Moscow: Foreign Languages Publishing House, 1946), pp. 16–17.

42. Aumont, *Montage Eisenstein,* p. 55.

43. Eisenstein, cited in ibid., p. 49.

44. Lev Kuleshov, "Americanitis," in *Kuleshov on Film,* ed. and trans. Ronald Levaco (Berkeley: University of California Press, 1974), p. 130.

45. Dziga Vertov, "From the History of the Kinoks," in *Kino-Eye: The Writings of Dziga Vertov,* ed. Annette Michelson, trans. Kevin O'Brien (Berkeley: University of California Press, 1984), pp. 98–99.

46. Dziga Vertov, "From Kino-Eye to Radio-Eye," in *Kino-Eye,* pp. 90–91.

47. Sergei Eisenstein, "The Cinematographic Principle and the Ideogram," in *Film Form,* ed. and trans. Jay Leyda (New York: Harcourt Brace Jovanovich, 1949), p. 41.

48. Sigfried Giedion, *Mechanization Takes Command* (New York: W. W. Norton, 1969; orig. pub. London: Oxford University Press, 1948), pp. 101–13.

49. Dziga Vertov, "Kino-Eye," in *Kino-Eye,* p. 67.

50. Dziga Vertov, "Kinoks: A Revolution," in *Kino-Eye,* p. 15.

51. Vertov, "From Kino-Eye to Radio-Eye," p. 88.

52. Dziga Vertov, "Kinopravda," in *Kino-Eye,* p. 132.

53. Vertov, "From Kino-Eye to Radio-Eye," p. 87.

54. A similar charge might be leveled at Noël Burch's distinction (discussed in chapter 1) between the "scientific" approach of the Lumière films and the artistic "illusionism" of supposedly more bourgeois representations. Even though Burch views the Lumière films as "scientific" on the basis of their visual polycentrism, rather than some notion of rationalized production, there seems to be little that is particularly "scientific" about that. Here, as with the avant-gardes, it seems that the recourse to a scientific model is based on visual criteria, on a simulation of the scientific: on the fact that Lumière's "scenes seem in fact to unfold before his camera rather like a microbic organism under a microscope, or like the movement of the stars seen through an astronomer's telescope" (Noël Burch, "Charles Baudelaire versus Doctor Frankenstein," *Afterimage,* nos. 8–9 [spring 1981]: 18).

55. These slogans are from Le Corbusier's *Towards a New Architecture,* p. 95. I should note that there are some exceptions to this rule; for example, most modernist prefabricated buildings and some modernist housing projects can, in fact, be seen as functional both in production and in their use.

56. Buckminster Fuller, cited in Banham, *Theory and Design in the First Machine Age,* p. 326.

57. Banham, *Theory and Design in the First Machine Age,* p. 162.

58. Benjamin, "Paris, Capital of the Nineteenth Century," p. 161.

59. Aumont, *Montage Eisenstein,* p. 56.

60. Jean Baudrillard, *For a Critique of the Political Economy of the Sign,* trans. Charles Levin (Saint Louis: Telos Press, 1981), p. 147.

4. Within the Space of High Tech

1. Bernard Stiegler, *Technics and Time,* vol. 1, *The Fault of Epimetheus,* trans. Richard Beardsworth and George Collins (Stanford, Calif.: Stanford University Press, 1998), p. 68.

2. Fredric Jameson, "The Cultural Logic of Late Capitalism," in *Postmodernism, or, The Cultural Logic of Late Capitalism* (Durham, N.C.: Duke University Press, 1991), p. 37.

3. Martin Heidegger, "The Turning," in *The Question concerning Technology and Other Essays,* trans. William Lovitt (New York: Harper Torchbooks, 1977), p. 48.

4. These ideas are presented in Heidegger's "The Question concerning Technology," in *The Question concerning Technology and Other Essays.* He also

discusses *technē* in "Building Dwelling Thinking," in *Poetry, Language, Thought,* trans. Albert Hofstadter (New York: Harper & Row, 1971).

5. Samuel Weber, "Theater, Technics, and Writing," *1-800* (fall 1989): 20 n. 8.

6. In using "aesthetic" here as a near synonym for "stylistic," I am suggesting that the notion of the "aesthetic," like that of technology, has changed. In this sense, "aesthetic" must be distinguished from the traditional notion of the aesthetic, which often, as in notions of the "beautiful," implies a sense of (organic) wholeness. Like the traditional notion of the aesthetic, however, it maintains the sense of a noninstrumental realm.

7. Joan Kron and Suzanne Slesin, *High Tech: The Industrial Style and Source Book for the Home* (New York: C. N. Potter, 1978), p. 1.

8. Clarence Barnhart, Sol Steinmetz, and Robert K. Barnhart, *The Second Barnhart Dictionary of New English* (New York: Harper & Row, 1980).

9. Kron and Slesin, *High Tech,* p. 1.

10. Kenneth Frampton, *Modern Architecture: A Critical History,* rev. and enl. ed. (New York: Thames & Hudson, 1985), p. 285.

11. James Botkin, Dan Dimancescu, and Ray Strata, with John McClellan, *Global Stakes: The Future of High Technology in America* (New York: Penguin Books, 1984), p. 20.

12. In common parlance, the term *black box* has come to be most readily associated with the instrument package designed to record flight data so that it can be recovered in the event of crashes. The use of the term in this sense, however, is merely an extension of its original usage.

13. André Bazin, "The Myth of Total Cinema," *What Is Cinema?* vol. 1, trans. Hugh Gray (Berkeley: University of California Press, 1967), pp. 17–22.

14. Thierry Chaput, "From Socrates to Intel: The Chaos of Microaesthetics," in *Design after Modernism: Beyond the Object,* ed. John Thackara (New York: Thames & Hudson, 1988), p. 185.

15. On "complexity," see Jean-François Lyotard, "Complexity and the Sublime," in *ICA Documents* 4–5 (1986): 10–12. On the relation of the techno-sciences to postmodernity, see in particular, Jean-François Lyotard, *The Postmodern Condition: A Report on Knowledge,* trans. Geoff Bennington and Brian Massumi (Minneapolis: University of Minnesota Press, 1984).

16. Jameson, "The Cultural Logic of Late Capitalism," p. 38.

17. Ibid. Indeed, Jameson has, in the notes to his book on postmodernism, regretted "the absence from this book of a chapter on cyberpunk," which he calls "the supreme *literary* expression if not of postmodernism, then of late capitalism itself" (p. 419).

18. The publisher's information for these books by Gibson is as follows: *Neuromancer* (New York: Ace Books, 1984), *Count Zero* (New York: Ace Books,

1986), *Mona Lisa Overdrive* (New York: Ace Books, 1988), and *Burning Chrome* (New York: Ace Books, 1986).

19. Gibson, *Neuromancer,* p. 3.

20. Kevin Kelly, "Cyberpunk Era: Interviews with William Gibson," *Whole Earth Review* 16 (summer 1989): 79.

21. Larry McCaffery, "An Interview with William Gibson," *Mississippi Review* 47–48(16) (2 and 3) (summer 1988): 229–30.

22. Kelly, "Cyberpunk Era," p. 79.

23. Bruce Sterling, "Preface," in *Mirrorshades: The Cyberpunk Anthology,* ed. Bruce Sterling (New York: Arbor House, 1986), p. ix.

24. Gibson, *Neuromancer,* p. 51.

25. Brooks Landon, *The Aesthetics of Ambivalence: Rethinking Science Fiction Film in the Age of Electronic (Re)production* (Westport, Conn.: Greenwood Press, 1992), p. 123.

26. Vivian Sobchack, *Screening Space: The American Science Fiction Film,* 2d enl. ed. (New York: Ungar, 1987), pp. 255–72.

27. Norman Spinrad, *Little Heroes* (New York: Bantam, 1987); John Shirley, *Eclipse* (New York: Blue Jay, 1985).

28. Sobchack, *Screening Space,* p. 248.

29. Claudia Springer, *Electronic Eros: Bodies and Desire in the Postindustrial Age* (Austin: University of Texas Press, 1996), p. 76.

30. Gibson, *Neuromancer,* p. 7.

31. Klaus Theweleit, *Male Fantasies,* 2 vols., trans. Stephen Conway, Erica Carter, and Chris Turner (Minneapolis: University of Minnesota Press, 1987, 1989). See, in particular, the first half of the chapter "Floods, Bodies, History," in volume 1, pp. 229–362.

32. Springer has already made this point quite explicit in her analysis of the hypermasculine cyborgs of recent sci-fi films, noting that "the Terminator and RoboCop display a rock-solid masculinity. Their technological adornments serve to heighten, not diminish, their bodies' status as fortresses. Cyborg imagery in the RoboCop and Terminator films exemplifies the invincible armored killing machine theorized by Klaus Theweleit in *Male Fantasies.* . . . Cyborgs like the Terminator and RoboCop realize the Freikorps fantasy in the realm of representation, making possible in fiction what can be only fantasized in fact" (*Electronic Eros,* p. 109).

33. Theweleit notes that "The unconscious is a flow and a desiring-machine" and argues that "It should again be stressed that it was not until late bourgeois society developed a notion of culture centering on the sharply delineated 'individual' that the machine and streams were set in opposition to the human body" (*Male Fantasies,* vol. 1, pp. 255–56).

34. Nicola Nixon, "Cyberpunk: Preparing the Ground for Revolution or Keeping the Boys Satisfied?" *Science-Fiction Studies* 19 (1992): 226.

35. Springer, *Electronic Eros*, 59.

36. Eva Cherniavsky, "(En)gendering Cyberspace in *Neuromancer*: Postmodern Subjectivity and Virtual Motherhood," *Genders* 18 (winter 1993): 38.

37. Clearly, the connection between the matrix and the maternal recalls Roger Dadoun's analysis of Metropolis as the "mother city," discussed in chapter 2. Fears of the matrix, and of the tech-noir city, as a dangerous, female fluidity also echo similar fears of the False Maria and the floods that she provokes. Yet, cyberpunk attitudes toward this technological fluidity and otherness are not, especially in Gibson's case, simply a matter of fear.

38. Gibson, *Count Zero*, p. 119.

39. Gibson's tendency to represent complex, "unthinkable" technologies and technological spaces as feminine and "other" is confirmed in his novel *Idoru* (New York: G. P. Putnam's, 1996), whose story revolves around a famous musician's desire to marry a female Japanese pop idol who also happens to be an artificial intelligence—one who "dreams" her own videos; she is, apparently, an AI with an unconscious (indeed, she is described as a kind of "desiring-machine. . . . Not in any literal sense, . . . but please envision *aggregates of subjective desire*" [p. 178]). One character sensitive to these "unconscious," "aggregate" images describes her in terms that recall the overwhelming complexity of the matrix: "She is the tip of an iceberg, no, an Antarctica, of information. . . . she was some unthinkable volume of information" (ibid.).

40. Gibson, *Count Zero*, p. 120. The recounting of the story of "the Wig" is given on pages 120–23.

41. Of particular note here are N. Katherine Hayles's observations about "chaos theory" and gender. Noting that "Chaotic unpredictability and nonlinear thinking . . . are just the aspects of life that have tended to be culturally encoded as feminine," she reads the exclusion of women from James Gleick's book on the science of chaos (*Chaos: Making a New Science* [New York: Viking Press, 1987]) as a way of attaining "control over the polysemy of chaos, stripping it of its more dangerous and engendered aspects" (N. Katherine Hayles, *Chaos Bound: Orderly Disorder in Contemporary Literature and Science* [Ithaca, N.Y.: Cornell University Press, 1990], pp. 173–74).

42. Jameson, "The Cultural Logic of Late Capitalism," p. 37.

5. Technological Fetishism and the Techno-Cultural Unconscious

1. Karl Marx, *Capital* (New York: International Publishers, 1967), vol. 1, part 1, chap. 1, sec. 2–3.

2. Max Horkheimer and Theodor W. Adorno, *Dialectic of Enlightenment,* trans. John Cumming (New York: Continuum, 1972).

3. Herbert Marcuse, *One-Dimensional Man: Studies in the Ideology of Advanced Industrial Society* (Boston: Beacon Press, 1964), p. 153.

4. Sigmund Freud, "The 'Uncanny,'" in *The Standard Edition of the Complete Psychological Works of Sigmund Freud,* vol. 17 (London: Hogarth Press, 1900), pp. 217–52.

5. A partial list might include rebellious robots or cyborgs (such as those depicted in Karel Capek's *R.U.R.*, Edgar Rice Burroughs's *Synthetic Men of Mars* as well as the more recent cyborg figures of *Blade Runner* [1982], *The Terminator* [1984], and *Eve of Destruction* [1990]), computers that turn against their human creators (as in *The Forbin Project* [1970], *2001: A Space Odyssey* [1968], *Demon Seed* [1977], or *Tron* [1982]), or those cases where technology or its by-products provoke mutations or other destructive havoc [e.g., *Dr. Cyclops* [1940], *Godzilla* [1954; U.S. release, 1956], *Them!* [1954], *The Amazing Colossal Man* [1957], *The Fly* [1958 and 1986], and *Mutant* [1983]).

6. Although he does not discuss *Metropolis,* Peter Wollen has noted the importance of orientalism to early modernism's rejection of instrumental reason. His analysis therefore seems relevant to my discussion of *Metropolis* here, and to my more general argument about the role of cultural otherness in modernism and postmodern techno-culture. See his chapter "Out of the Past: Fashion/Orientalism/The Body," in *Raiding the Icebox: Reflections on Twentieth-Century Culture* (Bloomington: Indiana University Press, 1993), pp. 1–34.

7. As a water monster, the Hydra also seems to fit nicely with *Metropolis*'s connection of a libidinal technology with the floods (which occur as a result of the False Maria's actions) that threaten to inundate the city. These floods, which rise up from beneath the city, seem also to represent unconscious urges. On this point, it is worth noting that Robert Graves cites the following explanation of the myth of the Hydra: "the Hydra was a source of underground rivers which used to burst out and inundate the land: if one of its numerous channels were blocked, the water broke through elsewhere" (Robert Graves, *The Greek Myths,* vol. 2 [Harmondsworth, England: Penguin Books, 1960], p. 109).

8. Anne Righter, "Introduction," in *The Tempest* (New York: Penguin Books, 1968), p. 42.

9. Octave Mannoni, *Prospero and Caliban: The Psychology of Colonization* (New York: Praeger, 1964), p. 109. Mannoni observes: "What the colonial in common with Prospero lacks, is awareness of the world of Others, a world in which Others have to be respected" (p. 108).

10. Isaac Asimov, *I, Robot* (New York: Ballantine Books, 1983 [1950]).

11. See Donna Haraway, "The Promises of Monsters: A Regenerative Politics

for Inappropriate/d Others," in *Cultural Studies,* ed. Lawrence Grossberg, Cary Nelson, and Paula Treichler (New York: Routledge, 1992), pp. 295–337.

12. The explanations of Godzilla's origins have not been consistent in each film of the series.

13. According to Chia-Ning Kao in *The Godzilla FAQ* (Frequently Asked Questions), "Kaiju is translated literally to 'mysterious beast/animal.' This is somewhat like monster but with a supernatural overtone. Japanese monster movies are often called 'kaiju eiga.' This is used for ALL Japanese monsters (even Gamera)." The Godzilla FAQ can be found at Barry's Temple of Godzilla, http://www.stomptokyo.com/godzillatemple/moreinfo.htm (March 1999) and elsewhere on the World Wide Web.

14. It is perhaps worth remembering at this point that J. Robert Oppenheimer, upon witnessing the first atomic explosion at Alamogordo, also had recourse to non-Western religious or supernatural discourse to describe atomic technology, quoting from the *Bhagavad Gita*: "I am become Death, the shatterer of worlds." Here, atomic technology seems figured as the ultimate example of a "technological sublime." Donna Haraway has in fact argued that this supernatural destructive power "forced humans to recognize their problematic kinship with each other as fragile earthlings at a scale of shared vulnerablity and mortality . . . explicitly ritualized at Alamogordo" by Oppenheimer's words (Donna J. Haraway, *Modest Witness@Second Millennium.FemaleMan© Meets OncoMouse™ Feminism and Technoscience* [New York: Routledge, 1997], p. 55).

15. An interesting contrast to *Akira* can be found in the monstrous technological "incorporation" represented in Shinya Tsukamoto's 1989 film *Tetsuo: The Iron Man.* The film, as Tony Rayns has observed, uses "organic metal as a metaphor for subconscious fears and libidinal urges." Thus, technology seems in *Tetsuo* to take on a supernatural, and certainly fetishistic, life of its own. See Tony Rayns, "Tokyo Stories," *Sight and Sound* ns. 1.8 (August 1991): 15.

16. Donna Haraway, "A Manifesto for Cyborgs: Science, Technology and Socialist Feminism in the 1980s," *Socialist Review* 80 (1985): 65–108; Gilles Deleuze and Félix Guattari, *A Thousand Plateaus: Capitalism and Schizophrenia,* trans. Brian Massumi (Minneapolis: University of Minnesota Press, 1987); Avital Ronell, *The Telephone Book: Technology, Schizophrenia, Electric Speech* (Lincoln: University of Nebraska Press, 1989).

17. On this distinction, see, for example, Kenneth Frampton, *Modern Architecture: A Critical History,* rev. enl. ed. (London: Thames & Hudson, 1985), p. 116, where he discusses the "tectonic" and "atectonic" impulses of modern architecture. See also my discussion in chapters 2 and 3.

18. Sherry Turkle, *Life on the Screen: Identity in the Age of the Internet* (New York: Simon and Schuster, 1995), p. 166.

19. Michael Bremer, *SimLife User Manual* (Orinda, Calif.: Maxis, 1992), p. 6; cited in ibid., p. 168.

20. Steven Levy, *Artificial Life: The Quest for a New Frontier* (New York: Pantheon Books, 1992), pp. 7–8.

21. The idea of a complex system is therefore similar in some ways to the idea of a "black box," referring to any design component whose internal workings cannot, or need not, be known. The importance of the "black-box" metaphor to high-tech aesthetics is discussed in chapter 4.

22. Manuel DeLanda makes this point in his intriguing examination of complexity, self-organizing processes, and the idea of nonorganic life ("Non-organic Life," in *Incorporations*, ed. Jonathan Crary and Sanford Kwinter [New York: Zone Books, 1992], pp. 129–67).

23. Vernor Vinge, "True Names," in *True Names and Other Dangers* (New York: Baen Books, 1987), p. 81.

24. Marge Piercy, *He, She and It* (New York: Ballantine Books, 1991), pp. 24–25. Piercy's novel is in fact a rewriting of the myth of Rabbi Loew's creation of a Golem, and focuses much of its attention on the question of artificial intelligence's ability to develop its own agency, as opposed to being merely a tool for humans. It is perhaps worth noting also that Piercy acknowledges both William Gibson and Donna Haraway as influences on her own work.

25. Scott Bukatman, *Terminal Identity: The Virtual Subject in Postmodern Science Fiction* (Durham, N.C.: Duke University Press, 1993); see, in particular, pp. 201–5.

26. Mark Dery, *Escape Velocity: Cyberculture at the End of the Century* (New York: Grove Press, 1996), p. 278.

27. Sigmund Freud, "Animism, Magic, and the Omnipotence of Thoughts," in *Totem and Taboo*, in *The Standard Edition of the Complete Psychological Works of Sigmund Freud*, vol. 13 (London: Hogarth Press, 1900), p. 88.

28. Claudia Springer, *Electronic Eros: Bodies and Desire in the Postindustrial Age* (Austin: University of Texas Press, 1996), pp. 58–59; cited on pp. 209–10 of Bukatman, *Terminal Identity*.

29. Cited in Erik Davis, "Technopagans," *Wired* 3.07 (July 1995): 176.

30. John Perry Barlow, "Africa Rising," *Wired* 6.01 (January 1998): 142–58.

31. "*Playboy* Interview: Marshall McLuhan," *Playboy* (March 1969): 64.

32. Davis, "Technopagans," p. 126–33, 174–81. Davis has more recently published a book that deals more broadly with the connections between technology and "spirituality": Erik Davis, *Techgnosis: Myth, Magic and Mysticism in the Age of Information* (New York: Harmony Books, 1998).

33. Teshome Gabriel has drawn attention to the "connection"—and, indeed, relevance—of such "other" discourses to the world of computers and high tech

in his exploration of the debt that digital technologies owe to the practice of traditional weaving in Africa and other non-Western societies, and how weaving offers a more suggestive, "a-rational" model for conceptualizing digital technologies (Teshome H. Gabriel and Fabian Wagmeister, "Notes on Weavin' Digital: T(h)inkers at the Loom," *Social Identities* 3[3] [1997]: 333–43). Along similar lines, Sadie Plant has discussed the relevance of women and weaving to computer technologies (Sadie Plant, "The Future Looms: Weaving Women and Cybernetics," in *Clicking In: Hot Links to a Digital Culture,* ed. Lynn Hershman Leeson [Seattle: Bay Press, 1996], pp. 123–35). Plant has expanded on these ideas in her exceptional book *Zeros + Ones: Digital Women and the New Technoculture* (New York: Doubleday, 1997), where she emphasizes the many links between women, women's writing, and information technologies, drawing on a wide variety of sources, from French feminist thought to science-fiction novels to the diaries of Ada Lovelace.

34. See, for example, Paulina Borsook's sidebar article to Erik Davis's essay "The Goddess in Every Woman's Machine," *Wired* 3.07 (July 1995): 133, 181. Also see her humorous depiction of dealing with the "editorial boykings" of *Wired* in "The Memoirs of a Token: An Aging Berkeley Feminist Examines *Wired,*" in *Wired Women: Gender and New Realities in Cyberspace,* ed. Lynn Cherny and Elizabeth Reba Weise (Seattle: Seal Press, 1996), pp. 24–41. Borsook notes that she had in fact pitched the idea of an article on techno-pagans before Davis had, but *Wired* apparently only took the idea seriously when it was proposed by a "hip" male writer.

35. Haraway, "The Promises of Monsters," p. 299.

36. Martin Heidegger, "The Question concerning Technology," in *The Question concerning Technology and Other Essays,* trans. William Lovitt (New York: Harper Torchbooks, 1977), p. 10.

37. William Gibson, *Count Zero* (New York: Ace Books, 1986), p. 217.

38. Cited in Hari Kunzru, "You Are Borg," *Wired* 5.02 (February 1997): 157.

39. The three novels of Octavia Butler's "Xenogenesis" series are *Dawn* (1987), *Adulthood Rites* (1988), and *Imago* (1989), all published by Popular Library.

40. Donna J. Haraway, "The Biopolitics of Postmodern Bodies: Constitutions of Self in Immune System Discourse," in *Simians, Cyborgs, and Women: The Reinvention of Nature* (New York: Routledge, 1991), p. 228.

41. The Borg originally appeared in the second season of *Star Trek: The Next Generation* (episode 42, "Q Who?" May 6, 1989). They returned in a two-part episode spanning the hiatus between the third and fourth seasons (episodes 74 and 75, "The Best of Both Worlds," June 26, 1990, and September 24, 1990); in this episode, they abduct and "assimilate" Captain Picard, turning him into Locutus of Borg. They also appeared in episode 123 (May 18, 1992), titled "I, Borg," and

in episodes 152 and 153, "Descent," (June 21, 1993, and September 20, 1993). More recently they were the villains in the feature film *Star Trek: First Contact* (1996). Cf. Claudia Springer's discussion of the Borg in her *Electronic Eros,* pp. 59–61. There, Springer reads the Borg as a "matriarchal hive society" in contrast to the "patriarchal model" of human society on board the *Enterprise,* an interpretation that finds further support in the feature film *Star Trek: First Contact,* where the Borg are "headed" by a female "Queen Bee." On the origins of the "Queen Bee" idea, see the discussion in Edward Gross and Mark A. Altman, *Captain's Logs: The Unauthorized Complete Trek Voyages* (Boston: Little, Brown, 1995), pp. 199–200.

42. The Medusa figure has, of course, often served as a representation of the "monstrosity" and dangerous "otherness" of female sexuality, which must not be seen. Interestingly, however, Butler also describes the Oankali as "hydra-like." As I noted in my discussion of the False Maria's dance in *Metropolis,* the Hydra is yet another monster that has connections to female sexuality, water, and the unconscious. See note 7 above.

43. Richard Dawkins, *The Selfish Gene* (New York: Oxford University Press, 1976), p. 206. His explanation of memes takes place in chapter 11, "Memes: The New Replicators," pp. 203–15.

44. On the relation of memory and time to technology, see Bernard Stiegler, *Technics and Time,* vol. 1, *The Fault of Epimetheus,* trans. Richard Bearsdworth and George Collins (Stanford, Calif.: Stanford University Press, 1998).

45. Haraway, *Modest Witness@Second Millennium,* p. 136. Haraway's critique of the fetishism of genetics is elaborated in the chapter titled "Gene: Maps and Portraits of Life Itself," pp. 131–72.

46. John Plunkett and Louis Rossetto, eds., *Mind Grenades: Manifestos from the Future* (San Francisco: HardWired Books, 1996); Bill Gates, with Nathan Myhrvold and Peter Rinearson, *The Road Ahead* (New York: Viking Press, 1995). Again, see Borsook, "The Memoirs of a Token," for an account of the masculinist tendencies of the *Wired* editorial board.

47. This is is not to say that the fears of being "incorporated" within a technological or techno-cultural unconscious do not continue to haunt representations of technology in an age of high tech, as is illustrated not only in the example of the Borg, but in many contemporary science-fiction films, from the biotechnological threat of the monstrous aliens in the *Alien* series to the frightening fluidity and violent invasiveness of the T-1000 in *Terminator 2.*

Index

Created by Eileen Quam and Theresa Wolner

R. L. Rutsky is assistant professor of film, television, and theater at the University of Notre Dame. He is the author of numerous articles in *New German Critique, Film Quarterly, Cinema Journal, Strategies,* and *SubStance,* among other journals.